CARTOGRAPHICS
Designing the Modern Map

CARTOGRAPHICS : Designing the Modern Map
© 2016 SendPoints Publishing Co., Ltd.

EDITED & PUBLISHED BY SendPoints Publishing Co., Ltd.
PUBLISHER: Lin Gengli
PUBLISHING DIRECTOR: Lin Shijian
CHIEF EDITOR: Lin Shijian
EXECUTIVE EDITOR: Huang ShaoJun
ART DIRECTOR: He Wanling
EXECUTIVE ART EDITOR: Peng Peng
PROOFREADING: Sundae Li, Ellen Christensen

REGISTERED ADDRESS: Room 15A Block 9 Tsui Chuk Garden, Wong Tai Sin, Kowloon, Hong Kong
TEL: +852-35832323 / **FAX:** +852-35832448
OFFICE ADDRESS: 7F, 9th Anning Street, Jinsha Zhou, Baiyun District, Guangzhou, China
TEL: +86-20-89095121 / **FAX:** +86-20-89095206
BEIJING OFFICE: Room 107, Floor 1, Xiyingfang Alley, Ande Road, Dongcheng District, Beijing, China
TEL: +86-10-84139071 / **FAX:** +86-10-84139071
SHANGHAI OFFICE: Room 307, Building 1, Hong Qiang Creative Zhabei District, Shanghai, China
TEL: +86-21-63523469 / **FAX:** +86-21-63523469

SALES MANAGER: Sissi
TEL: +86-20-81007895
EMAIL: overseas01@sendpoints.cn
WEBSITE: www.sendpoints.cn / www.spbooks.cn

ISBN 978-988-14703-3-1
All rights reserved. No part of this publication may be reproduced, stored in a retrieval system or transmitted in any form or by any means, electronic, mechanical, photocopying, recording or otherwise, without prior permission in writing from the publisher. For more information, please contact SendPoints Publishing Co., Ltd.
Printed and bound in China.

PREFACE

Jasmine Desclaux-Salachas[1]

I am told there are people who do not care for maps, and find it hard to believe.

...And the map itself with its infinite, eloquent suggestion, made up the whole of my materials. It is perhaps not often that a map figures so largely in a tale; yet it is always important. The author must know his countryside whether real or imaginary, like his hand; the distances, the points of the compass, the place of the sun's rising, the behaviour of the moon, should all be beyond cavil.

Robert Louis Stevenson, spring 1893
My first book, Treasure Island (Tusitala Edition, vol. II, page XXV)

A gorgeous culmination of Stevenson's viewpoint from long ago, CARTOGRAPHICS: *Designing the Modern Map* ushers in a new and advanced understanding of mapping which links diverse disciplines through the use of observation, data, technological innovation, collage, and illustration.

Mapping means using many techniques to express what must be said, constructing a visionary and universal language. To achieve this aim, cartographers must keep readers in mind all the way when producing maps, deciding what information need to remain and what should be discarded on the map in order to create a universal language that makes sense to readers at first sight.

Maps are travel companions, keeping us linked to the space as we stroll through the world out of necessity or desire. They measure, trace, and show the world for us, enabling us to better understand the societies and the world in which we live—they function as a golden thread leading us through the maze of life. Anyone curious about our world can, through maps, find out detailed information and its origin, from a local to a global recognition, or can even listen to dreams.

Mapping is a complicated process. An accurate map always requires tons of work and efforts. It means data collection and documentation of their sources, in order to produce a set of information, which is then graphically drawn, allowing spontaneous interpretation by readers. From this perspective, a map is a communication tool. An honest work must begin with a serious survey, which should investigate the situation of the target territory, like the measurement of the field and analysis of its objects at a 1:1 scale, which specifically characterizes "civil mapping."

"Civilian mapping" is produced by completing the so-called "base map", implemented to outline a series of topographical maps generated from those measurements by "generalization." "Civilian mapping" enables an in-depth reading of the territories and their intermingling issues and offers true information to each citizen, whoever he is and wherever he is, within a defined perimeter.

As a state-of-the-art scientific tool helping us to make decisions, maps can be found everywhere in our daily lives, representing the world in its smallest details, in any form and at any scale.

Today, though our digital tools create the illusion that a map can be produced in a few clicks, or that a map is merely a quick illustration, designed on-the-go, maps can be complex, opinionated, political, or personal.

Maps are often handled by the traveller without any real awareness of the infinite care involved in their design and production: research, data acquisition or collection, methods to structure those data and the rigorous procedures involved in reproducing them. The *'Cafés-cartographiques'* association, founded in France, welcome the public into this universe usually known only by professionals, who have worked unsparingly for generations at the forefront of technical progress enriched by a variety of related fields.

Cartographers work alongside geographers, historians, researchers, and scientists[2] from a wide range of disciplines in order to help them better express their views. Cartographers also work with technicians who supply them with tools and reproduce their work, whatever the media of distribution.

It is urgent to provide opportunities for people simply interested in sharing knowledge and creativity to gather together in an informal setting focusing on mapping. Gatherings are expected to offer occasions for all the curious and amateurs to meet students and professionals and let them focus on the little-known works of qualified authors.

We may consider that cartography remains a subjective art, a historical medium for ideological domination, propaganda and conquest, but not so in its civilian expression. A map can be much more. It is not merely an image, but a processing of multi-disciplinary data synthesized for a wide audience for the common interest. When information is designed with rigor and honesty, we are rewarded with a beautiful outcome, intelligible to the public at large.

Maps are fantastic tools, enabling us to understand our world. Maps are able to federate citizenships, curiosities, sensibilities, and creativities. The following pages are a special invitation to enjoy this vision.

---

1 Jasmine Desclaux-Salachas: cartographer, cartography teacher, doctoral student and founder and director general of Les Cafés-cartographiques.

2 Les Cafés-cartographiques association is an organization founded by Jasmine D. Salachas in 1999 which focuses on cartography (The website is under construction). For more information please check https://www.facebook.com/CafesCartographiques/info/?tab=page_info.

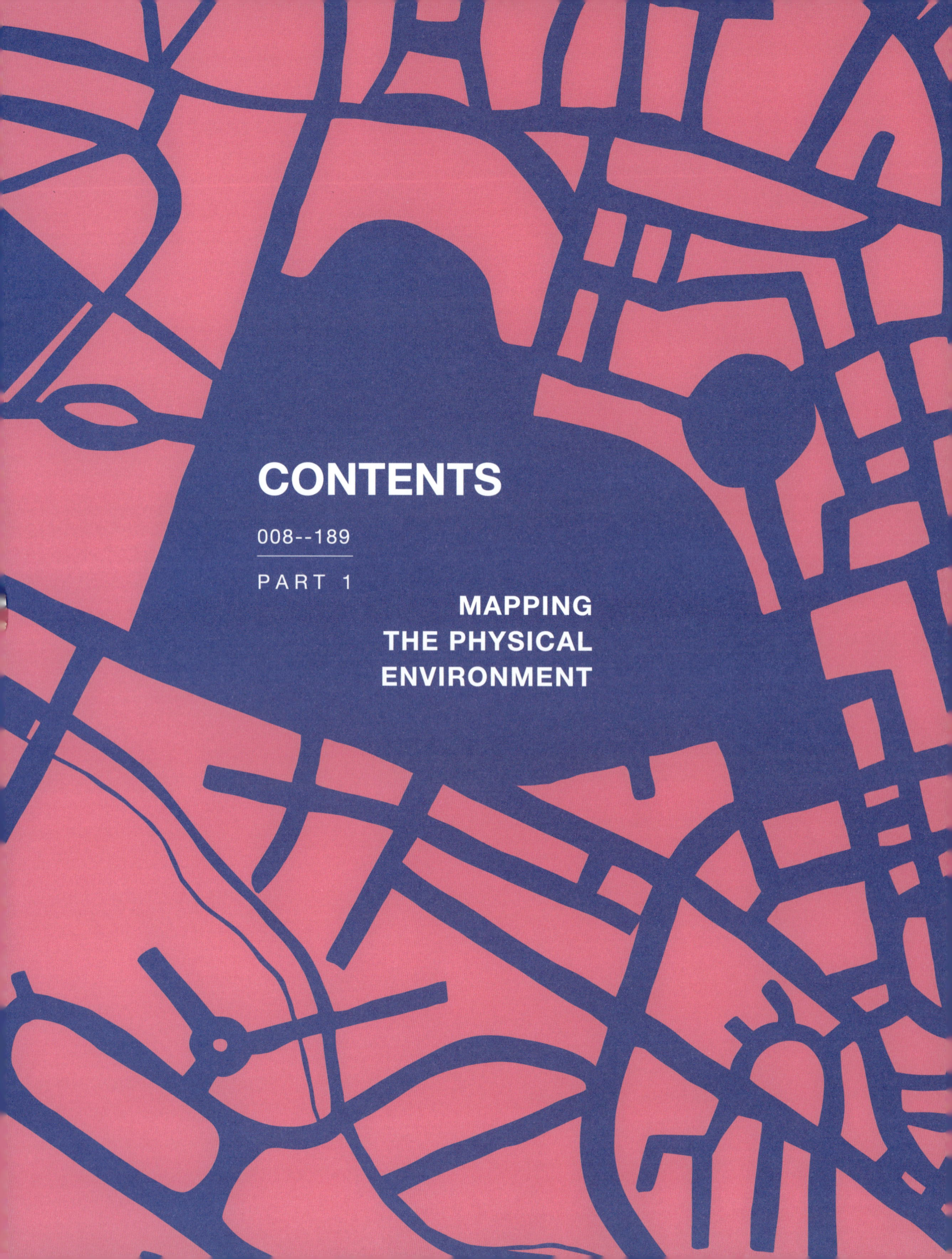

CONTENTS

008--189

PART 1

MAPPING
THE PHYSICAL
ENVIRONMENT

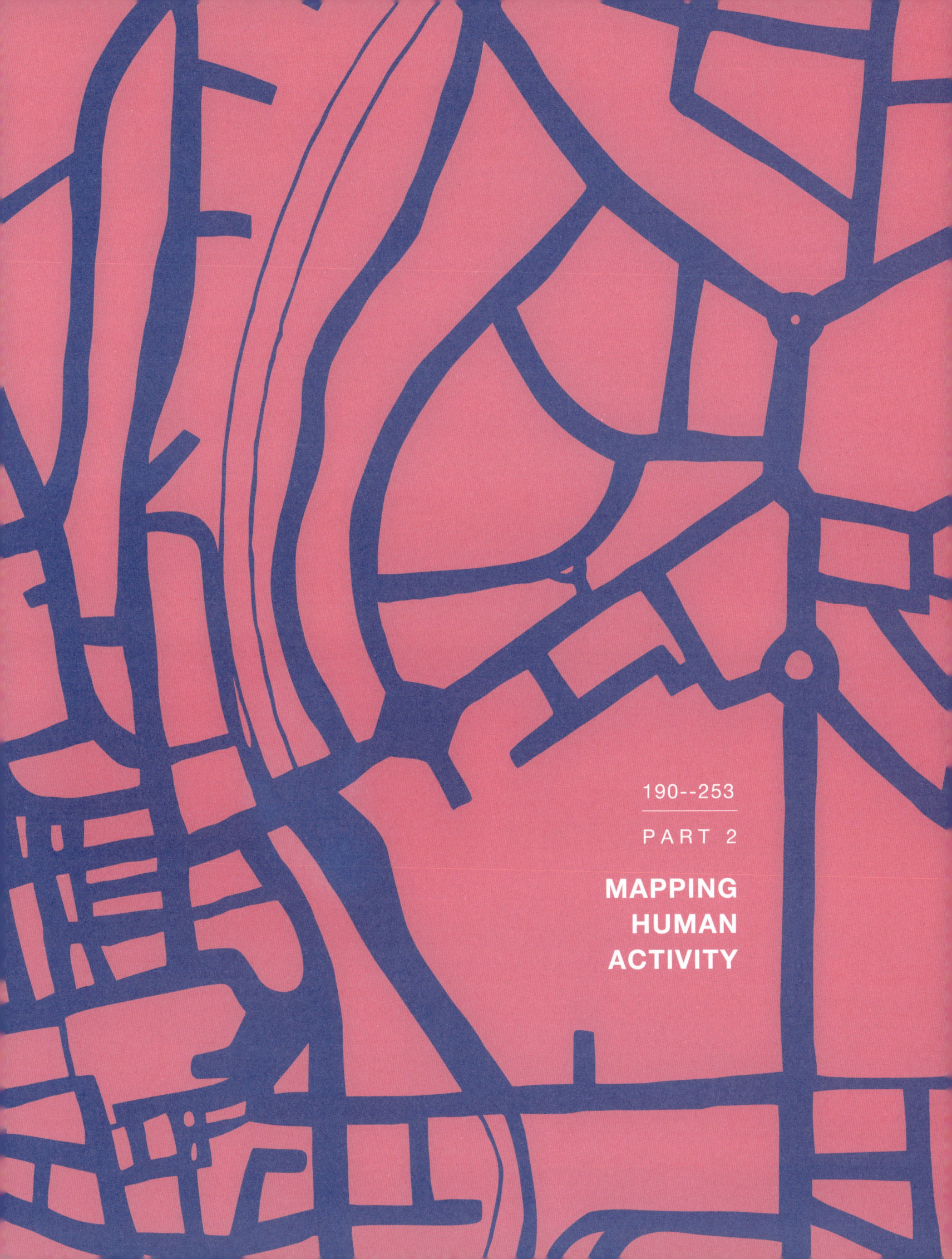

190--253

PART 2

**MAPPING
HUMAN
ACTIVITY**

PART 1

MAPPING THE PHYSICAL ENVIRONMENT

This section features maps that portray the physical environment of a geographic area at a particular point in time. Public transportation maps such as railway maps are included, as are fictional maps.

MAPPING THE PHYSICAL ENVIRONMENT

Egg Map

Egg map is a perfect gadget for people who want to avoid oversized printed maps or digital maps. It fits in your hand or your pocket and has manual zooming. Just squeezing it, you'll get more details about the place than bumbling with a printed map of your information source. The map ball is characterized by incredible flexibility and it is also lightweight. It's filled with 100% oxygen. To make it better, it's made of waterproof material so you can use it even in unfriendly weather conditions like rain, mud or snow. Each quarter has a different color so you can't take the wrong turn. Just keep zooming to find out more about sights, public transport or the nearest restaurant.

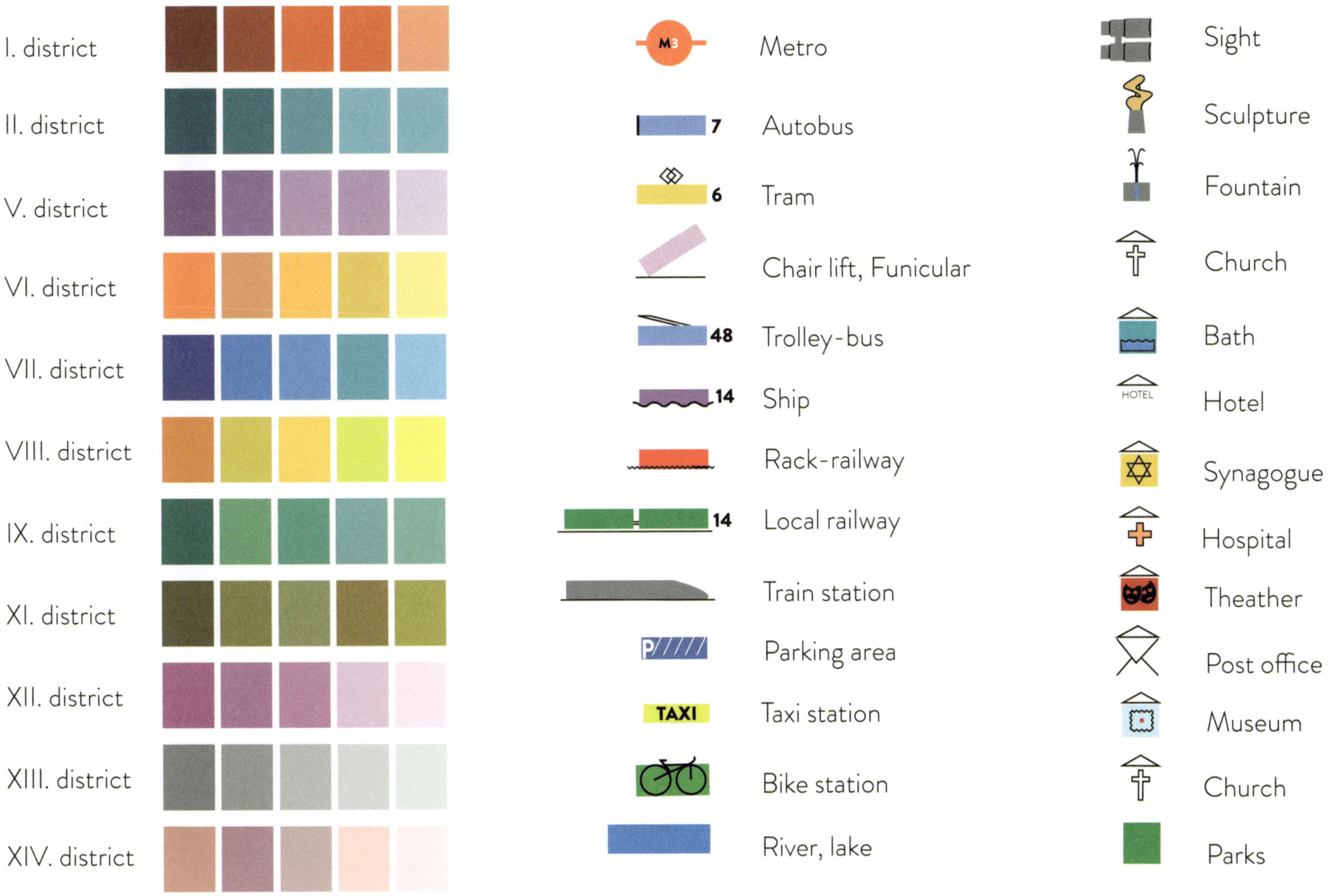

I. district						Metro		Sight
II. district					7	Autobus		Sculpture
V. district					6	Tram		Fountain
VI. district						Chair lift, Funicular		Church
VII. district					48	Trolley-bus		Bath
VIII. district					14	Ship	HOTEL	Hotel
IX. district						Rack-railway		Synagogue
XI. district					14	Local railway		Hospital
XII. district						Train station		Theather
XIII. district						Parking area		Post office
XIV. district					TAXI	Taxi station		Museum
						Bike station		Church
						River, lake		Parks

DESIGN Dénes Sátor COUNTRY/REGION Hungary

MAPPING THE PHYSICAL ENVIRONMENT

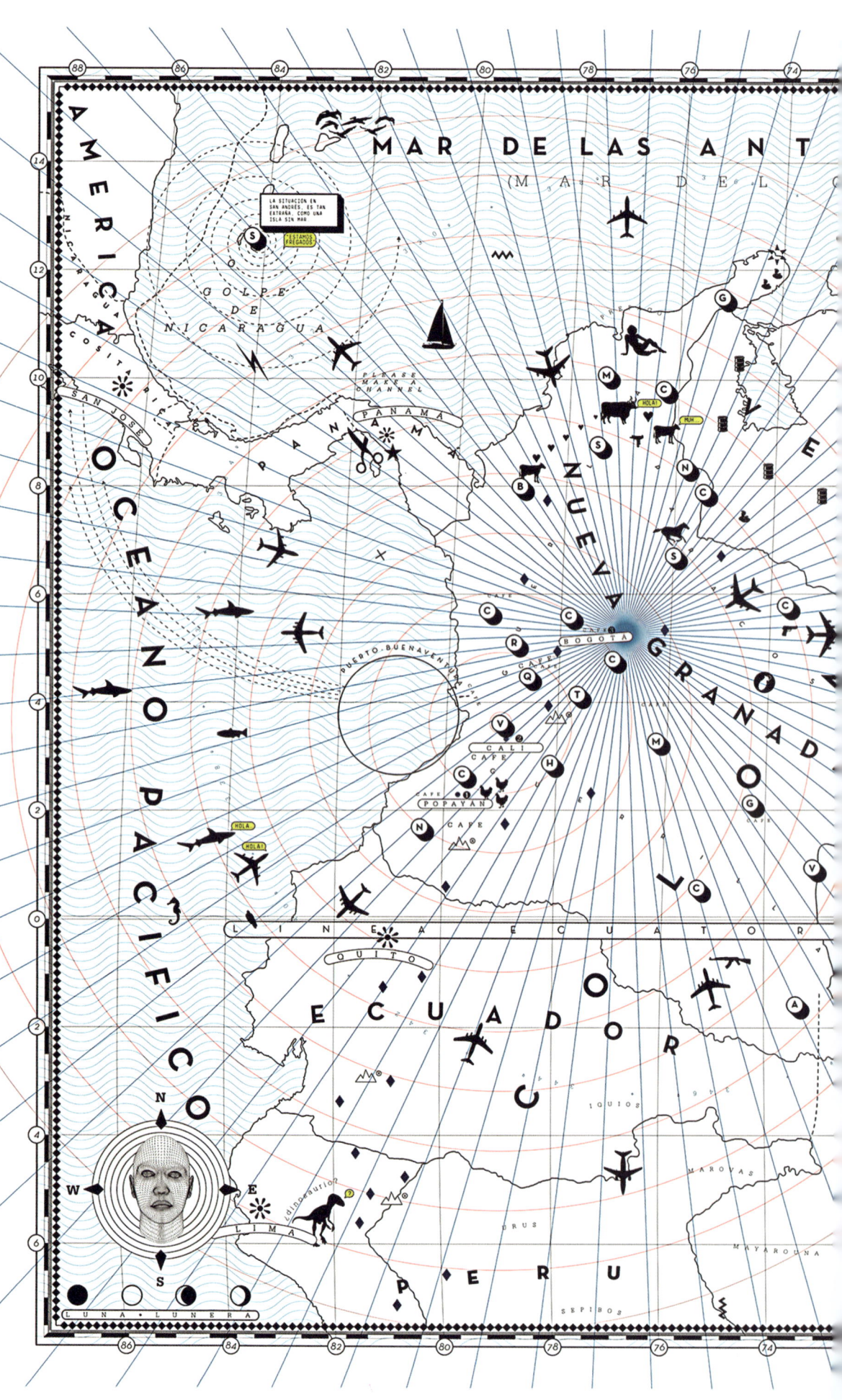

Unofficial Map

This map is based on a classic map of modern Colombia. It was redesigned to reflect the current geography of Colombia. Meanwhile, it is full of the designer's reflections and thoughts about North and South America. The map purposely reveals a territorial blur and makes fun of the actual circumstances. It dramatizes where conflicts took place and also highlights the attributes and negative aspects of every little place demarcated on the map. The use of different keys helped the construction of this map.

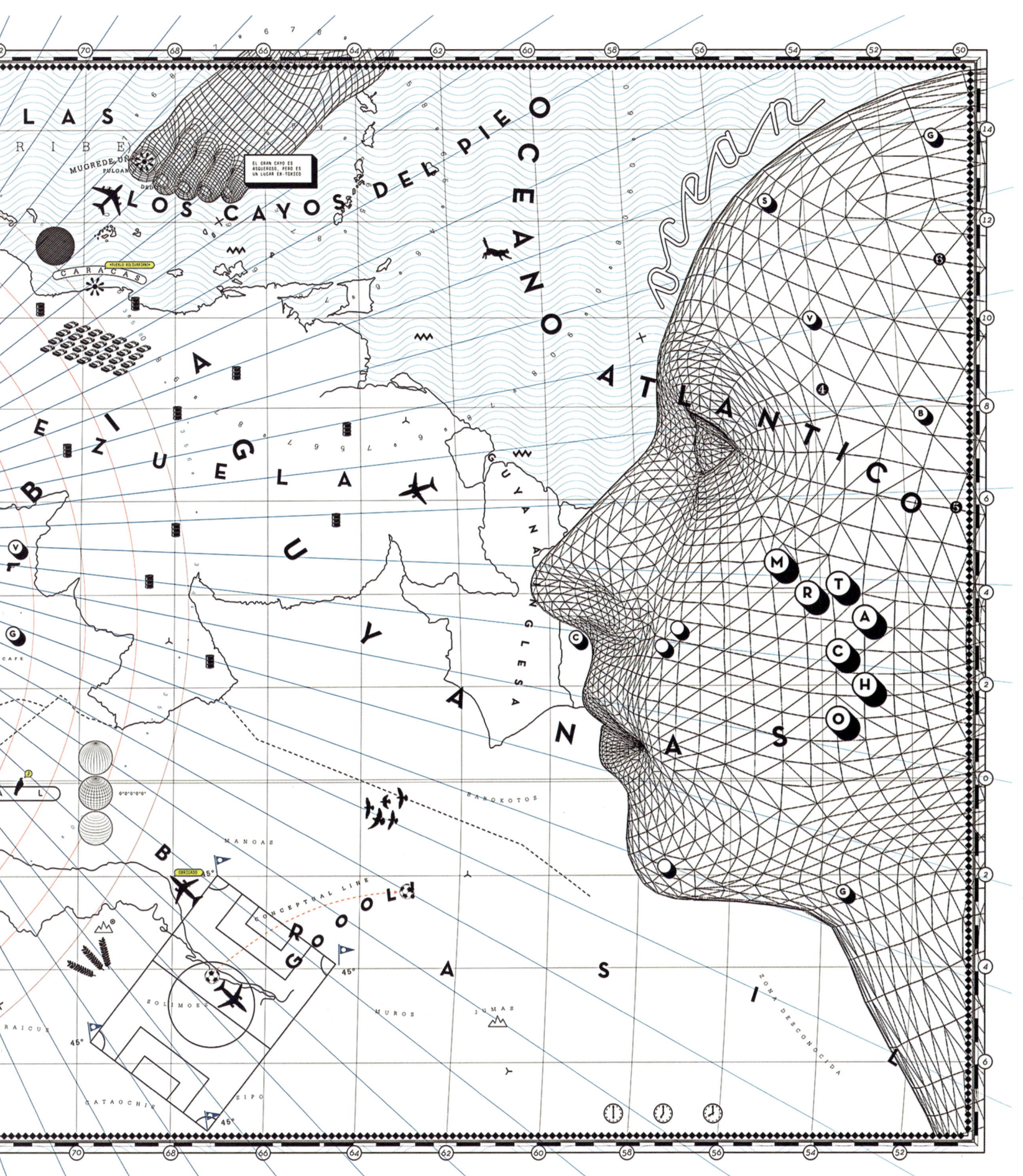

World Power Map

The map shows about 20,000 power plants of various types around the world, such as wind energy and solar power. The dataset was sourced from the site Enipedia and processed by Postgresql. The map is made using HTML5 and WebGL (Mapbox GL lib) and the basemap is OpenStreetMap rendered by CartoDB.

Hand Drawn Map of San Francisco

Commissioned by Evermade.com, the designer illustrated a complex map of San Francisco that showcased the city in all of its glory. The map was created after months of research. The main landmarks can be found on the map as well as some interesting facts, sub-cultures of the people in each area, good restaurants, shops and bars, etc. The map was created using a Wacom tablet and Photoshop, which means that it was editable.

DESIGN   Jenni Sparks   COUNTRY/REGION   The UK

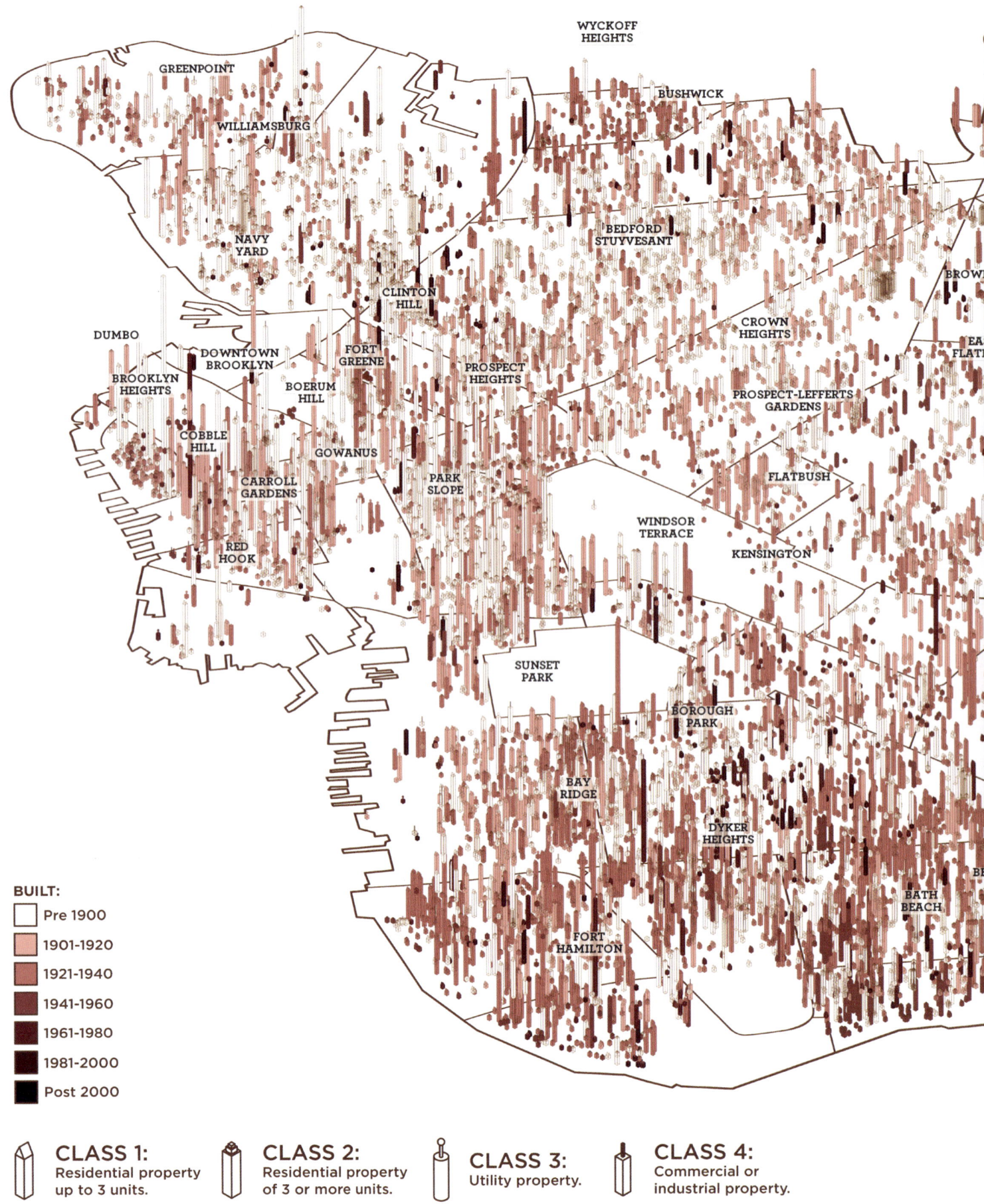

MAPPING THE PHYSICAL ENVIRONMENT

BUILT:
- Pre 1900
- 1901–1920
- 1921–1940
- 1941–1960
- 1961–1980
- 1981–2000
- Post 2000

CLASS 1: Residential property up to 3 units.

CLASS 2: Residential property of 3 or more units.

CLASS 3: Utility property.

CLASS 4: Commercial or industrial property.

Gentrification: Brooklyn

Gentrification: Brooklyn is a project that attempts to visualize the effects of gentrification up to the year 2010. By using the info provided by NYC Open Data, an interactive map was created, which focuses on real estate by mapping price by square foot, building class, and date of construction. The interactive map allows users to view the visualization in isometric or perspective view, as well as being able to view neighborhoods individually.

MAPPING THE PHYSICAL ENVIRONMENT

Reproduced with the permission of the British Geological Survey.
© NERC. All rights reserved.

Bicentennial Geological Map of Britain

This map commemorates the two-hundredth anniversary of the publication of the first geological map of Britain, by William Smith (1769-1839). It depicts modern geological bedrock data in the style of Smith's early nineteenth century cartography.

DESIGN   Craig Woodward   COUNTRY/REGION   The UK

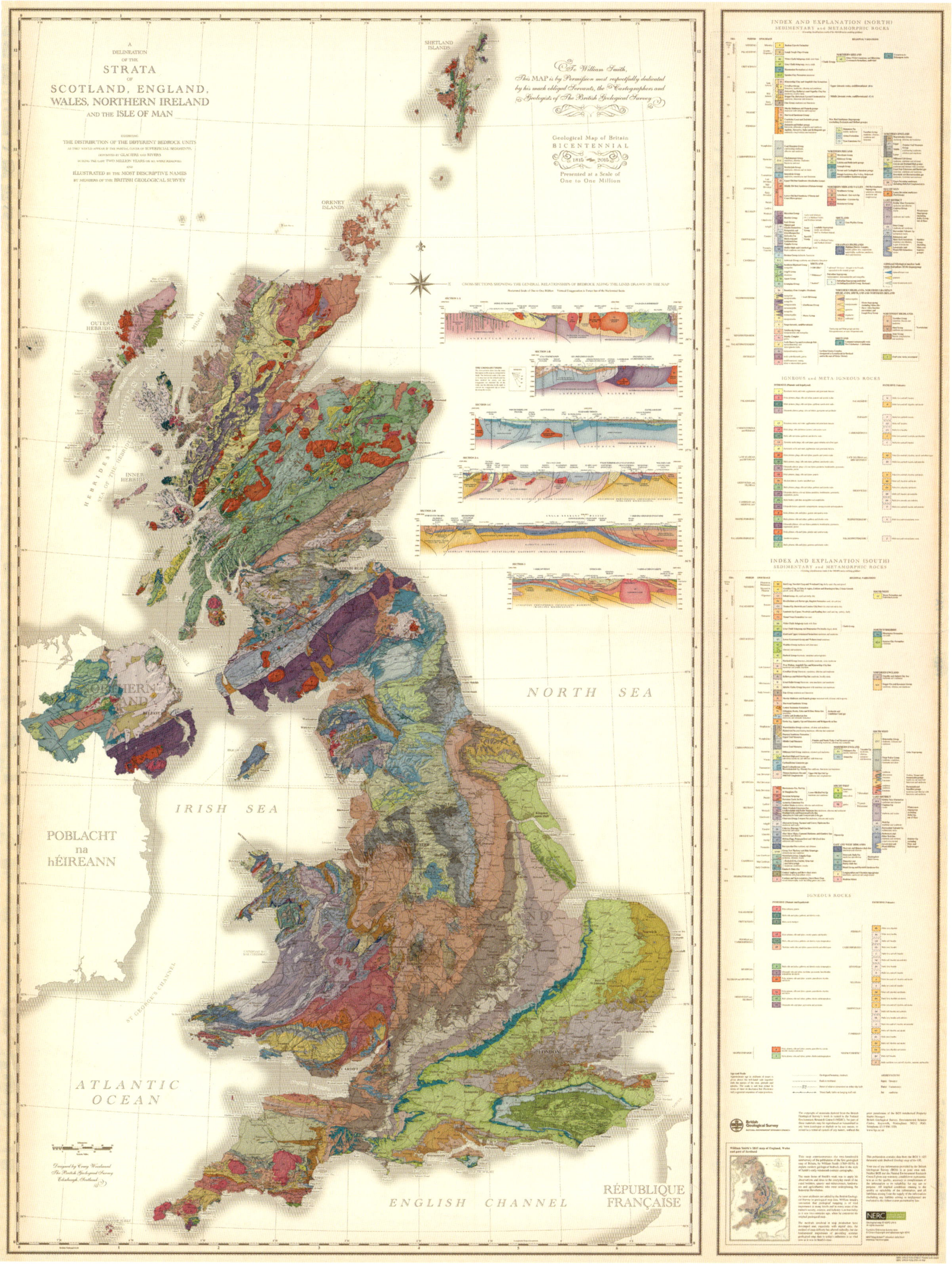

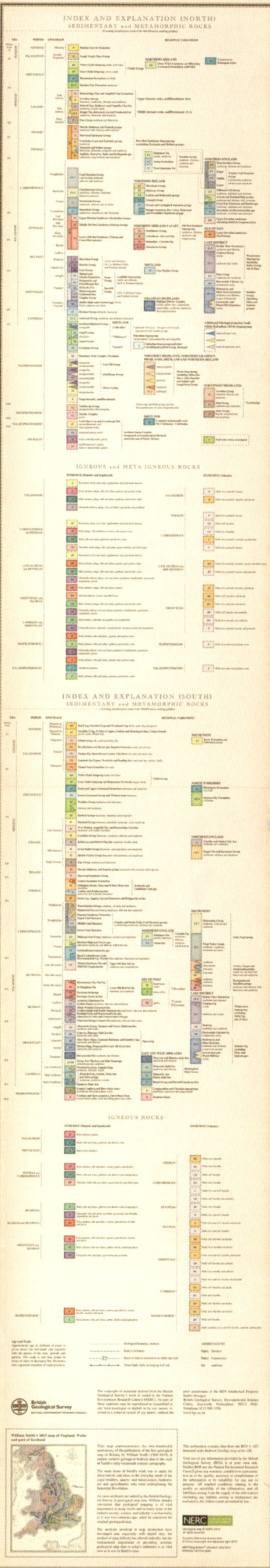

Literary London Map

This map of London features characters from the pages of novels based in the capital, including the famous and infamous, the less well known and also those with an amazing moniker or brilliantly conceived nickname who are a credit to their creator. Each character has been plotted in the corners of the city they most liked to roam or chose to call home (Sometimes at Majesty's Pleasure). Combining hand-drawn typography and illustration, more than 250 novels were mined in the making of this piece.

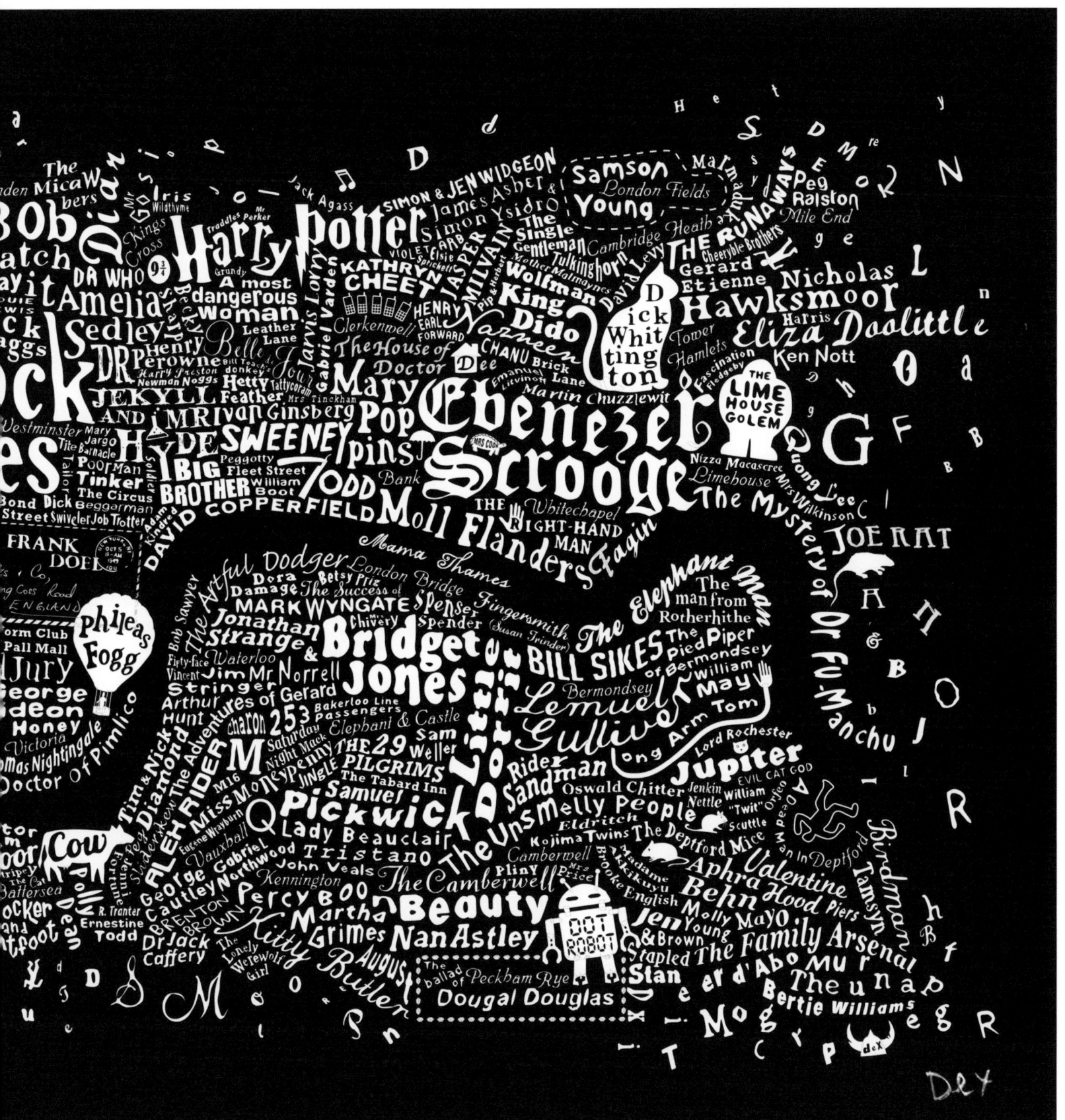

Hell USA Map

This map shows the geographic "Devils" and accompanying "Hells" across the country. The data was sourced by the designer himself by different search methods. Here are some notable findings: the more interesting landscapes of the mountains and coasts generally appear to earn the infernal titles. Incidentally, there's a surplus of Devil's Den's, Backbones, Elbows, Kitchens, and Hell Holes beside the more unique brands like Devil's Cup and Saucer Island, Devil's Icebox, and Satan's Kingdom. Another odd phenomenon is that none of the Apostle Islands bear apostle's names, yet a Devil's Island is included. Also, residential roads occasionally have names like Evil Lane.

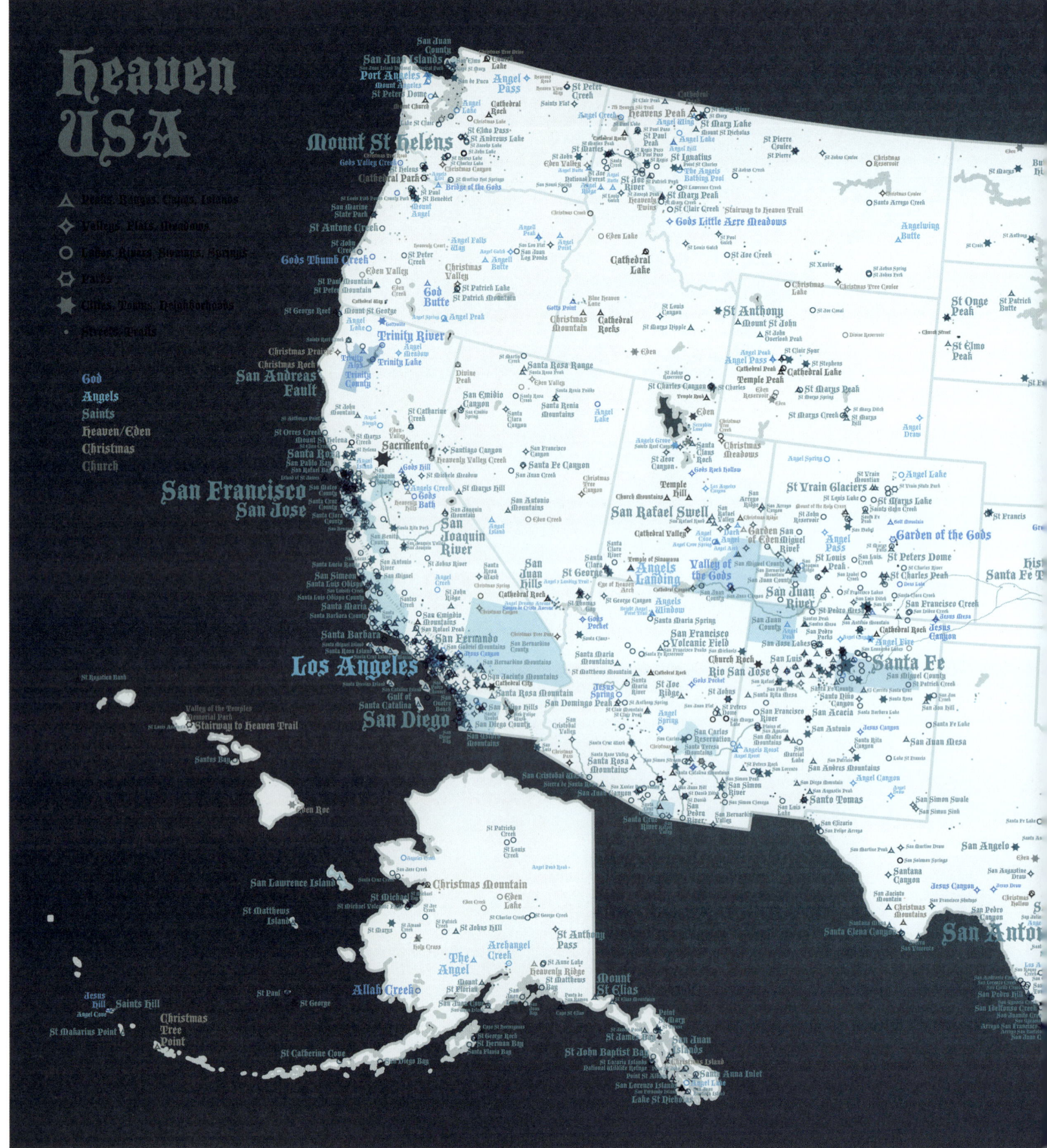

Heaven USA Map

A bit of Heaven on Earth in this graphic which maps place names using derivations of God, Heaven, Saint, Church, Angel, and Christmas. The populated areas often hold more of the heavenly names with particular concentrations in the areas colonized by the Spanish and French. Quirky names include God's Bath, Angel's Trumpet Road, Angel Tear Way, In God's Hand Way, and 6 Santa Claus Lanes. There are also intriguing titles as Bridge of the Gods across the Columbia, Stairway to Heaven Trail in Hawaii, Cathedral Caverns in Alabama, and Angel's Landing in Zion National Park.

Illustrated Island Maps

These are two maps of Zanzibar and Magadascar, and are part of a series of illustrated maps of islands. Although simplified and more decorative than practical, these maps are still geographically accurate, providing the most important geographical information such as cities, islands, capes, bays and places of interest. At the same time they are fantastic colorful worlds populated by local flora and fauna, such as chameleons, frogs, butterflies, and exotic plants.

DESIGN Bianca Tschaikner   COUNTRY/REGION Austria

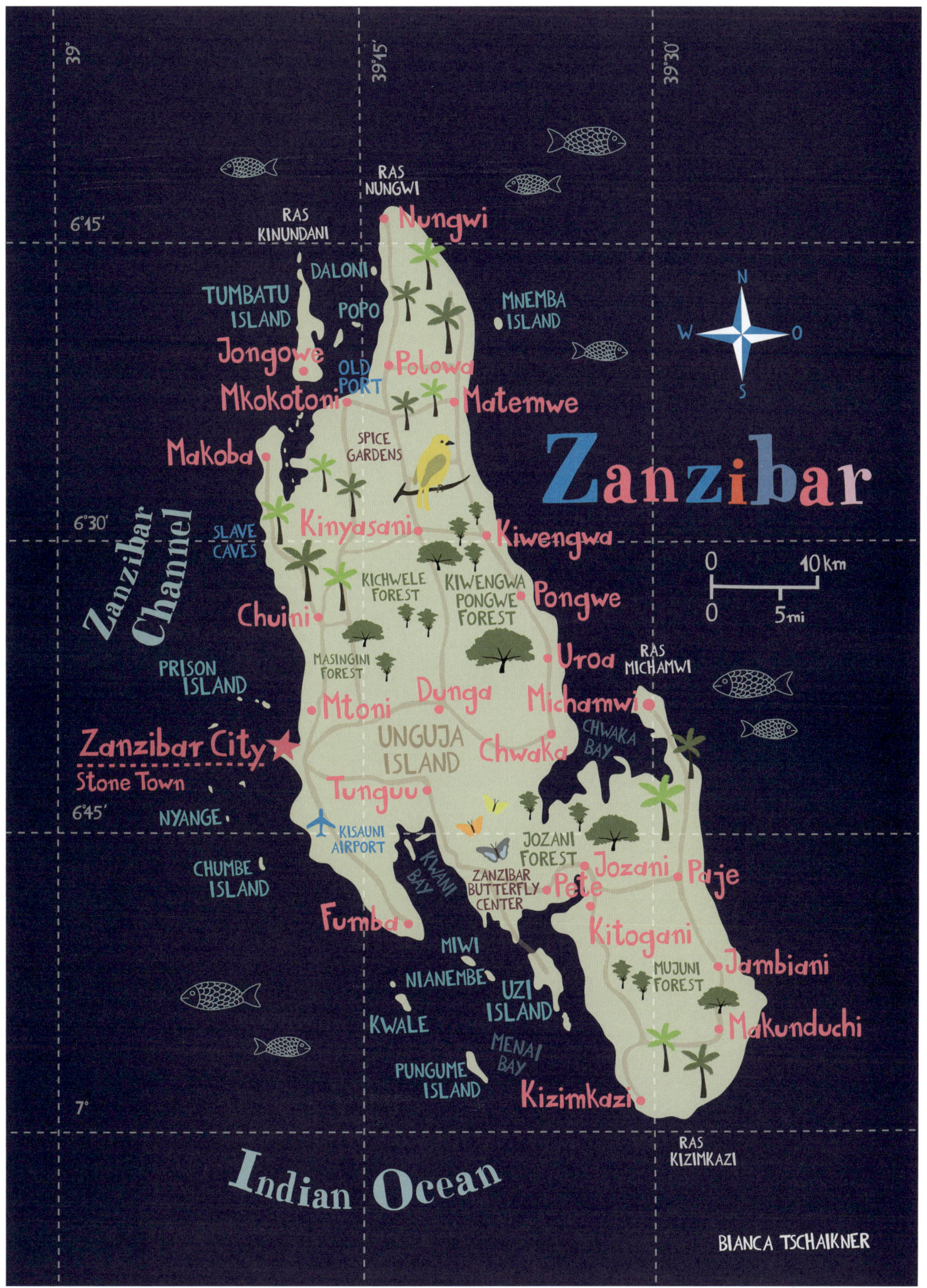

Map of The Hobbit

J.R.R. Tolkien's fiction inspired the creation of this map. It covers the area where The Hobbit takes place, from the Shire to the Lonely Mountain. The original size is 100x50cm.

Map Arda Tataya Randasse

In J. R. R. Tolkien's legendarium, Arda is the name given to the Earth in a period of prehistory, wherein the places mentioned in The Lord of the Rings and related material once existed.

The Map of Arda includes Aman, Númenor and Ennor in the second era. Emphasis has been placed on the Aman, location of the Valar seats, and the other important places during this era. This is the first such map for Arda, made on the basis of maps KWFonstad, R. Foster, Encyclopedia of Middle-earth and Tolkien's works, and with the help of specialists. It is distinguished by high accuracy, and artistic treatment of the topic.

The map was made in fountain pen and color pencil on paper. Its original size is 100x50cm.

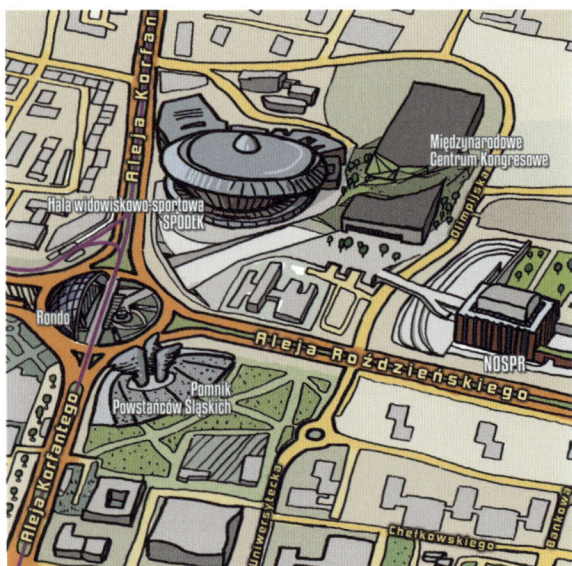
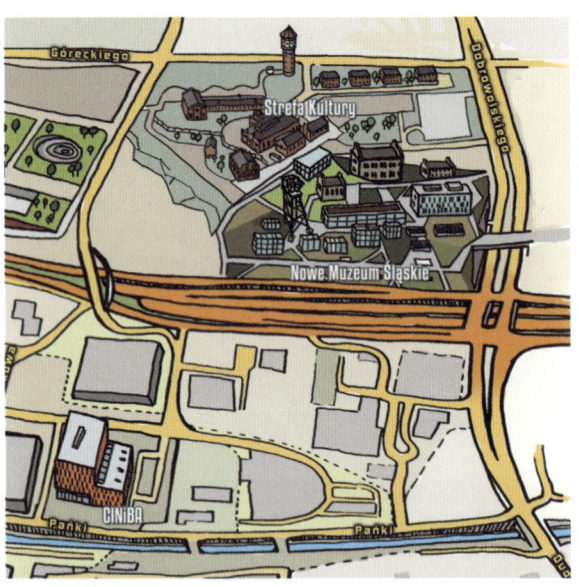

Map of Katowice City

This is a map of Katowice city drawn by tablet. In its center is a commercial district. The map shows some important buildings and objects in the downtown area. It could be printed in 50x70cm.

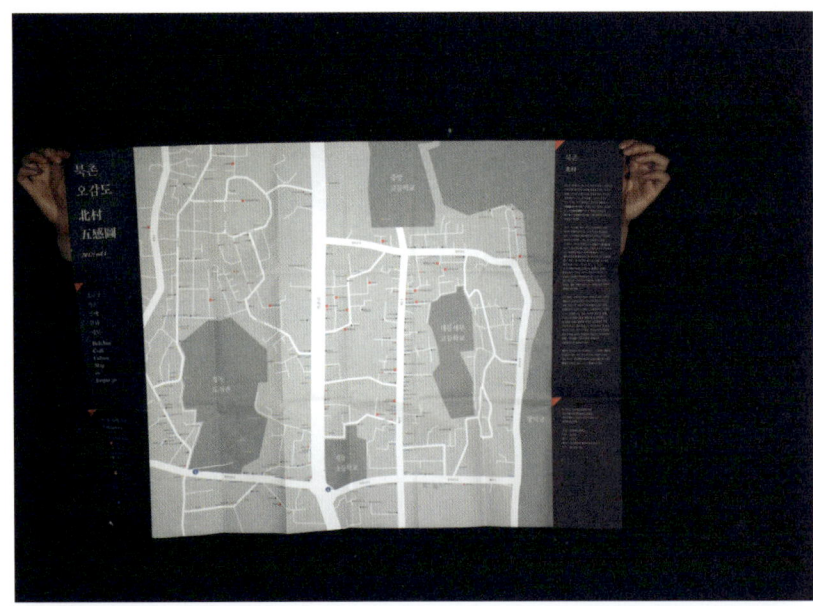

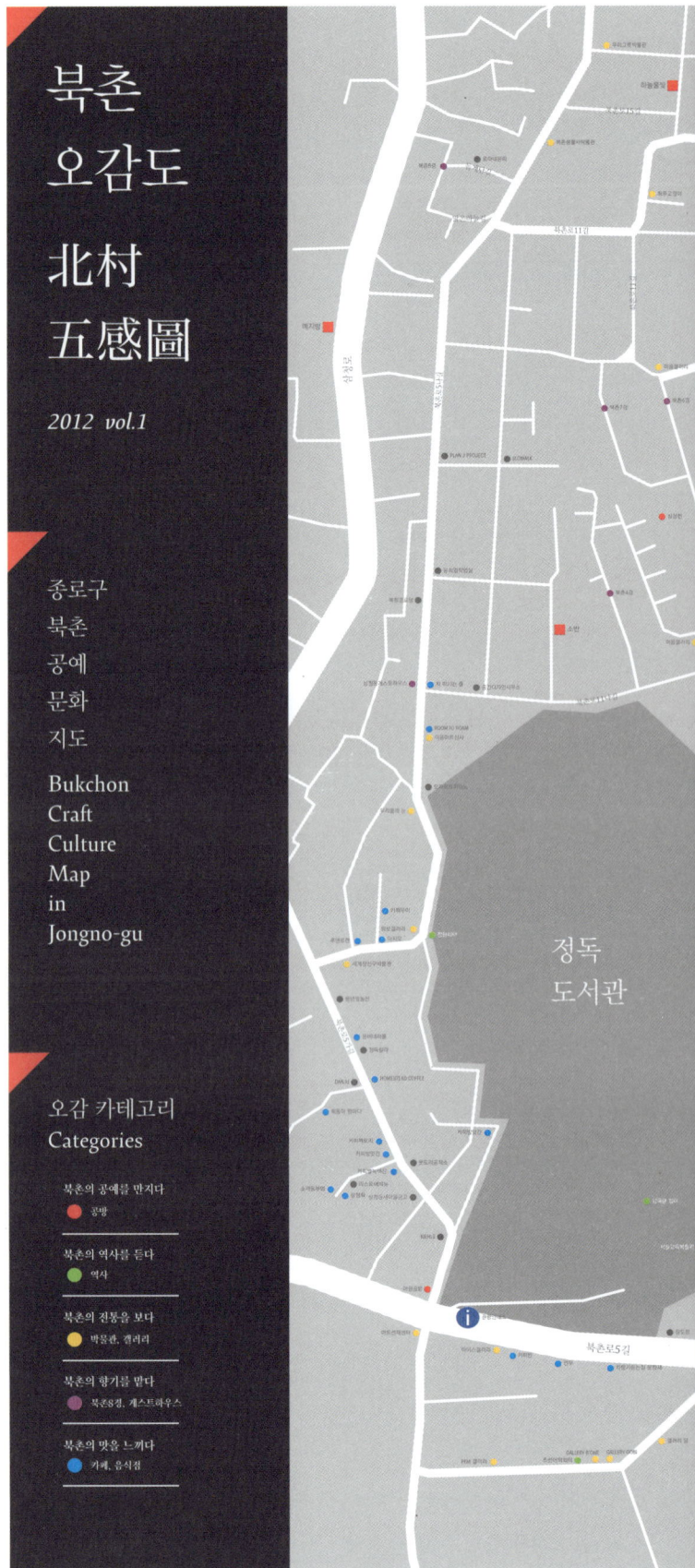

Bukchon Craft Culture Map in Jongno-gu

This is a map of Bukchon, Seoul, where Korean traditional houses are located. The concept was to focus on the remaining locations of traditional Korean craft and studios in Bukchon. The map recommends five paths for tourists to visit the main streets and subareas. In Bukchon, there are also various kinds of site-specific cafes, restaurants, historic places, and guest houses. They are categorized based on the responses of visitors to them, and the grade of importance and preference assigned by visitors. Given the scattered positions of these locations in Bukchon, a visual constellation idea was used. Five lines in different colors represent five suggested visit routes and their respective spots with descriptions. Moreover, QR codes are featured on the map for users to obtain information on the craft studios.

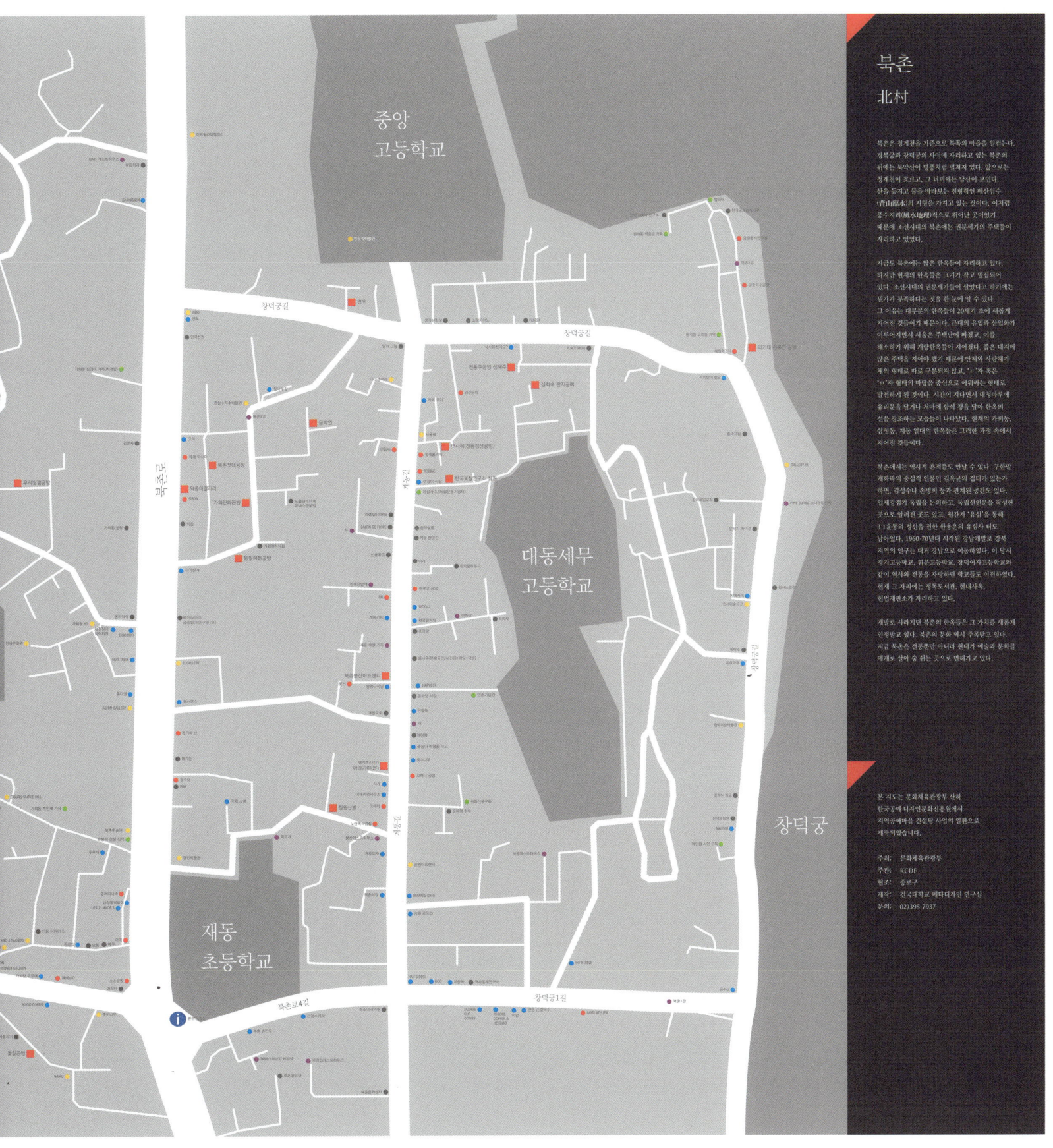

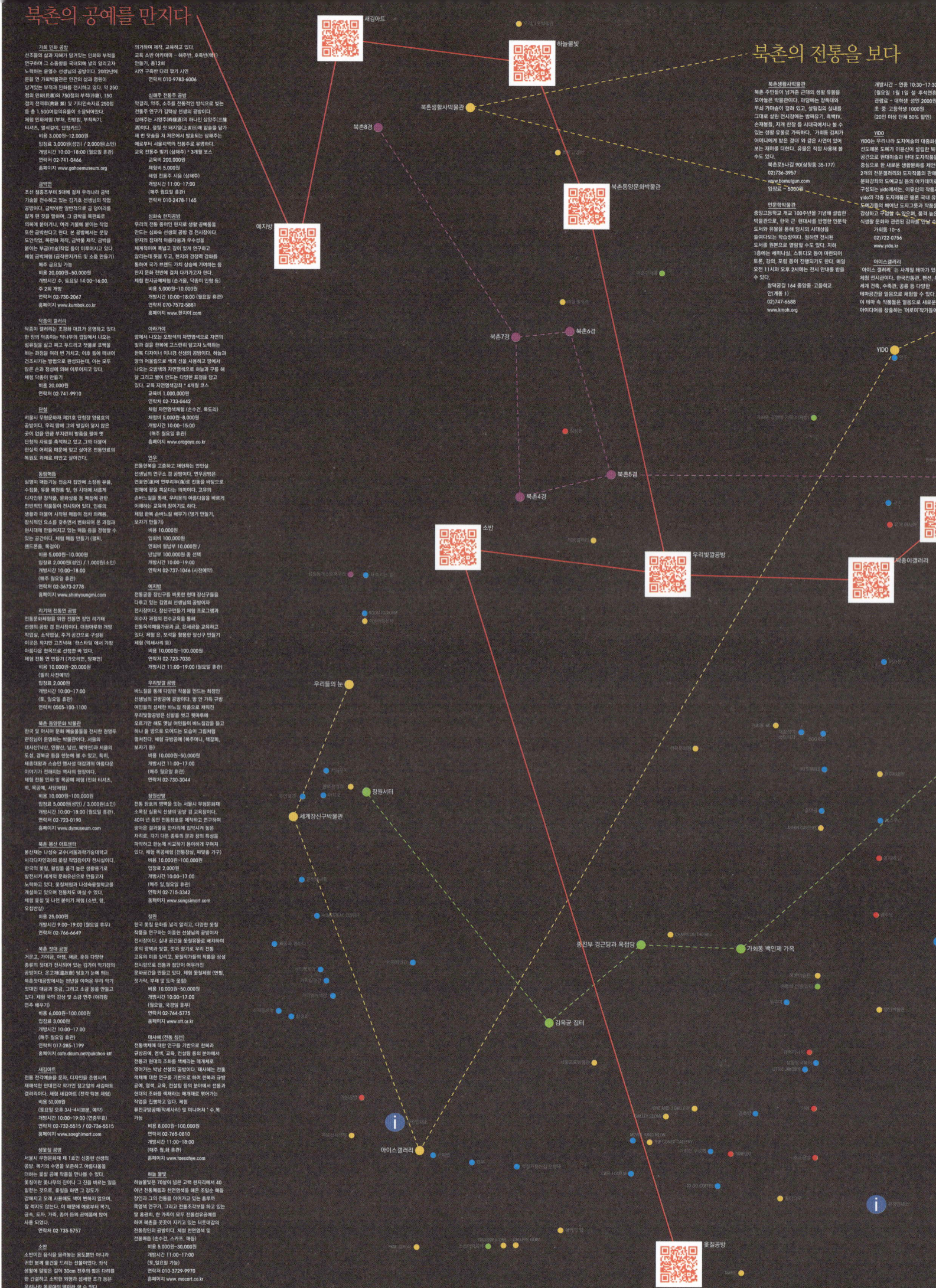

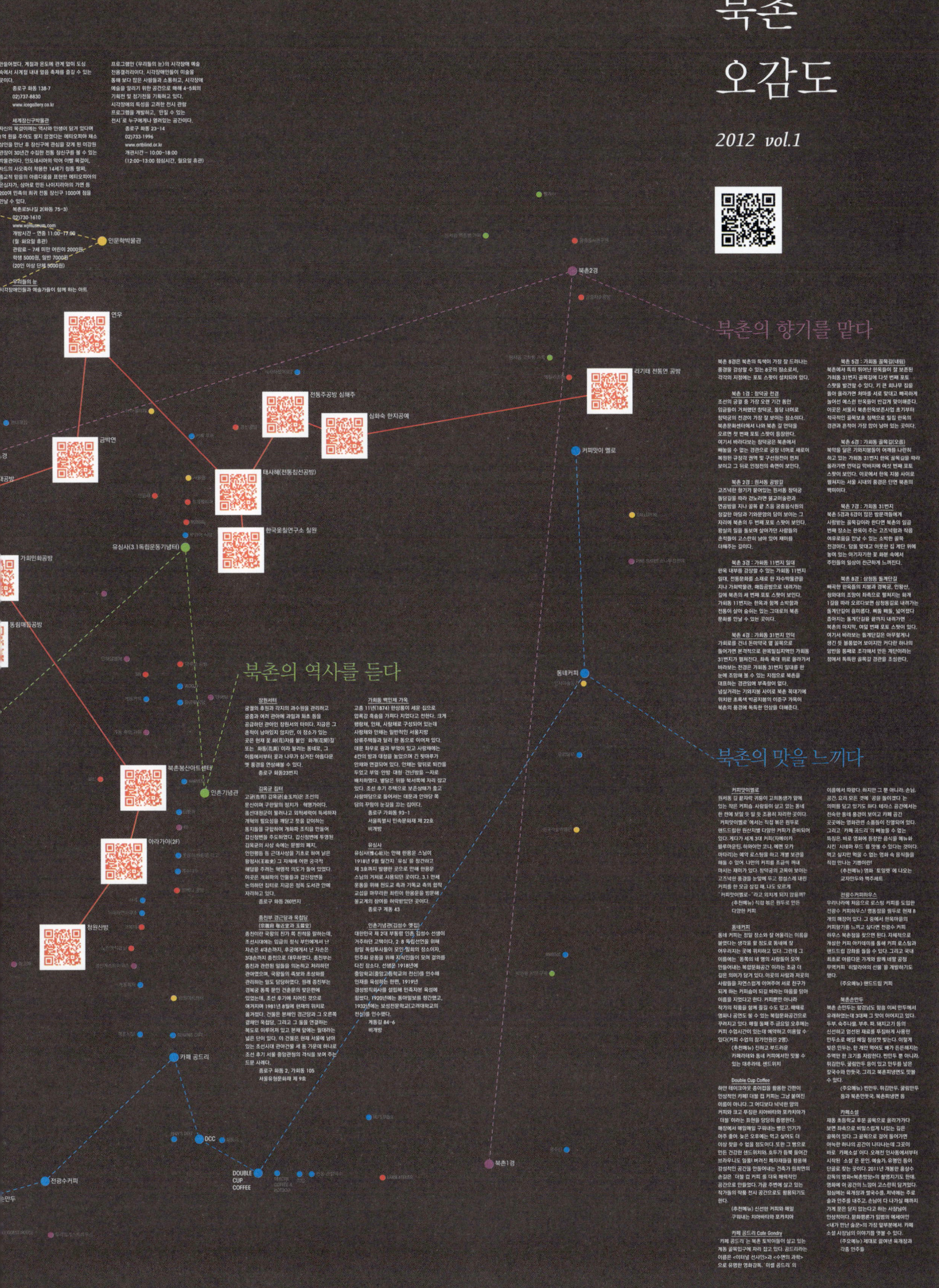

Prodotti Tradizionali Italiani

Prodotti Tradizionali Italiani is a large, detailed and colourful illustrated map of Italy and Italian food. It allows viewers to discover exhaustively all the wonders of Italian gastronomy, from the pecorino romano to the carciofi on gambo through delicious and tasty ascolana olives and hundreds of other specialties.

City Maps for Let's Go with Ryanair! Magazine

These illustrations of European cities are part of a series of illustrated city maps created for Let's Go with Ryanair! Magazine, the inflight magazine of Europe's number one discount airline RyanAir. They were created for the city guide section spotlight, which in the years 2012-2013 published a monthly portrait of a European city, consisting of a map and a guide with recommendations. Cities such as Edinburgh, Paris, Brussels, Porto, East London, Lisbon, Vienna, Rome and many more (All destinations of RyanAir) were featured in this section to provide inspiration for the reader's next trip and to help them to find their way around and to get to the most interesting places in town. The maps show restaurants, bars, coffee houses and a selection of sights, decorated with drawings of little scenes with people and buildings and to give an overall feel of the atmosphere of the city.

Wired World

Wired World was created as a donation for the Boise Weekly newspaper, located in Boise, Idaho. Made of oak and stained with English chestnut, the piece is composed of over 1,000 steel nails and roughly 150 feet of aluminum wire. The map aims to represent the infinite linkages within our world, regardless of physical or implied distance.

DESIGN Ciera Shaver   COUNTRY/REGION   The USA

Illustrated 3D Map of Australia (120cm X 120cm)

Illustrated 3D Maps

These two colorful maps are from Drakes's 3D map series. The maps are all hand sculpted and hand painted. The map base is built on foam board and the map is covered in a fine layer of papier mache, which is then painted, to form the background. The features on the map—including buildings and famous landmarks, food and wine, modes of transport and animals and plants are all handmade, carved and painted from a mixture of materials including balsa wood, beads, wire and acrylic paint. Each map is different and made to commission so features can be changed or added to personalize each piece.

Illustrated 3D Map of Italy (50cm X 70cm)

DESIGN Sara Drake   COUNTRY/REGION Australia

China Map Design

The map mainly presents the landmarks and major tourist attractions of each province in China. Leo put focus on the harmony of the whole map, insuring each illustration was uniform in style. Attention was also given to the color palette in order to enhance the end result.

Fuyang Eco Park Map of Animals and Plants

This is the distribution diagram of the animals and plants around the Eco-Pond in Fuyang Eco Park. The use of graphic illustrations makes the map distinctive from the old Eco maps.

DESIGN Wu, Jui-Che   COUNTRY/REGION Taiwan

Map of the Warnow River

The Warnow River flows in the north of Germany through Mecklenburg. The clichés associated with this area are the white beaches of the Baltic Sea, green cow pastures and fields of yellow blossoms. When drawing this map, Richter intended to avoid these stereotypes while still portraying the unspoilt nature along the banks of the river; shimmering color fields and the irregular forms of an ink drawing added to its natural look. In September, the land mass is the most beautiful—forests are still green, but the warm and harmonious shades of autumn are emerging everywhere. In contrast with the inland stand both the cool Baltic Sea and the city of Rostock, the biggest city in this area. Rostock marks the end of this quiet, nondescript river.

DESIGN Tilo Richter COUNTRY/REGION Germany

MAPPING THE PHYSICAL ENVIRONMENT

World Map Illustration

This map was not intended to be a very functional piece. Regarding the color palette, about 11 colors were used. Loosely speaking, the blue/grey colors denote capitalist countries like America while red colors denote the socialist ones like China. But it also attaches importance to matching the used colors. The typeface is meant to be more in a classic style that matches the 1950s or the 60s, so a Rockwell based font was used. The water pattern is based on a satellite image of the sky, so it's not a correct map revealing different depths of the sea; it is actually clouds in the water.

DESIGN Andreas Nilsson   COUNTRY/REGION Sweden   055

Food Map

Exploring new places through the food you eat is often a portal to the cultural complexities of that place. Inspired by a passion for travel, Caitlin Levin and Henry Hargreaves used many of the iconic foods of countries and continents and turned them into a series of physical maps, representing their interpretation of food from around the world. The maps show how food has traveled the globe—transforming and becoming a part of the cultural identity of that place. An example is that tomatoes originally came from the Andes in South America, but today Italy has become the tomato king. And people would think of Australia when "throw some shrimp on the barbie" was used. For those visiting France, it is a must to eat bread and cheese. These maps try to speak to the universality of how food unites people, brings people together and starts conversation.

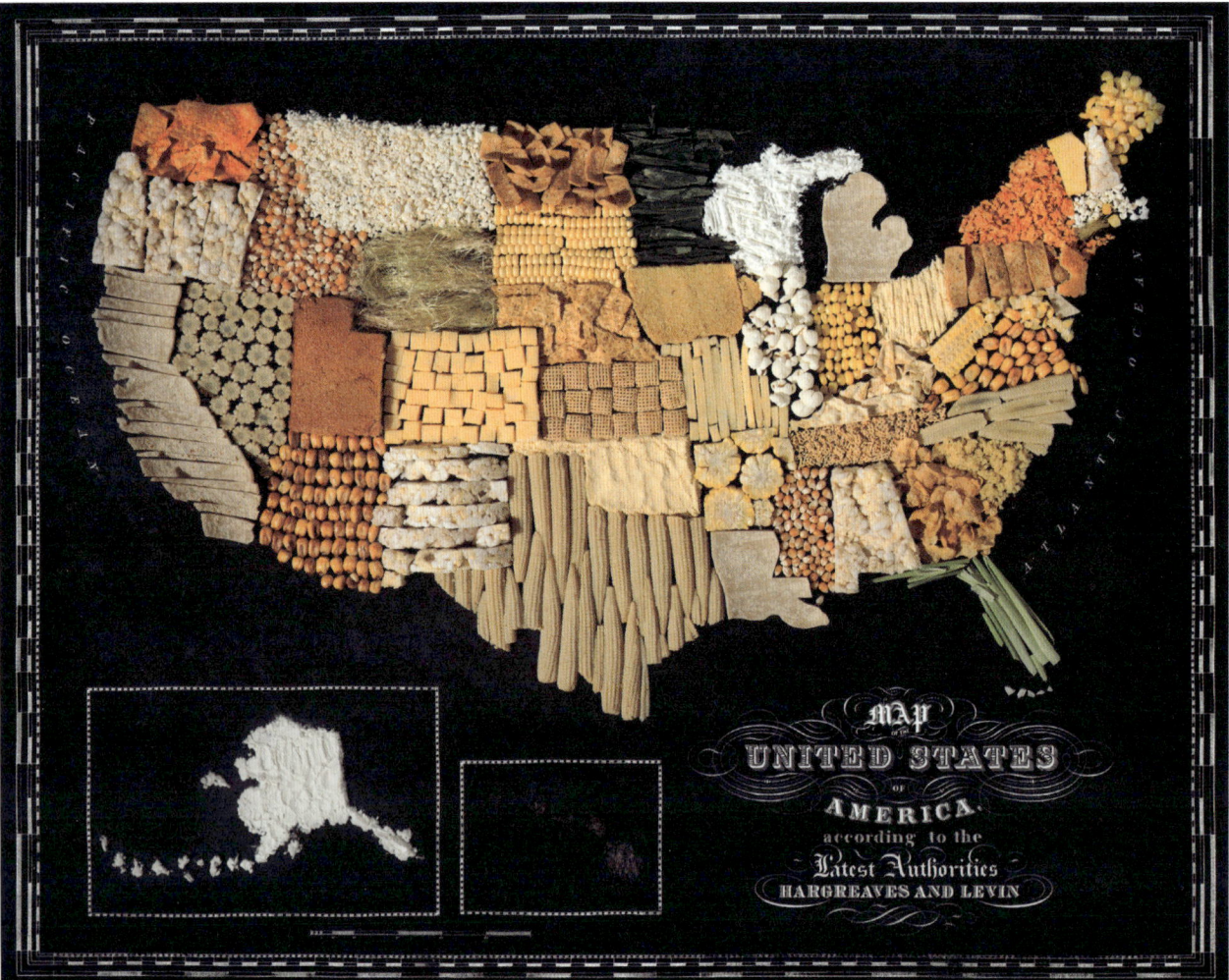

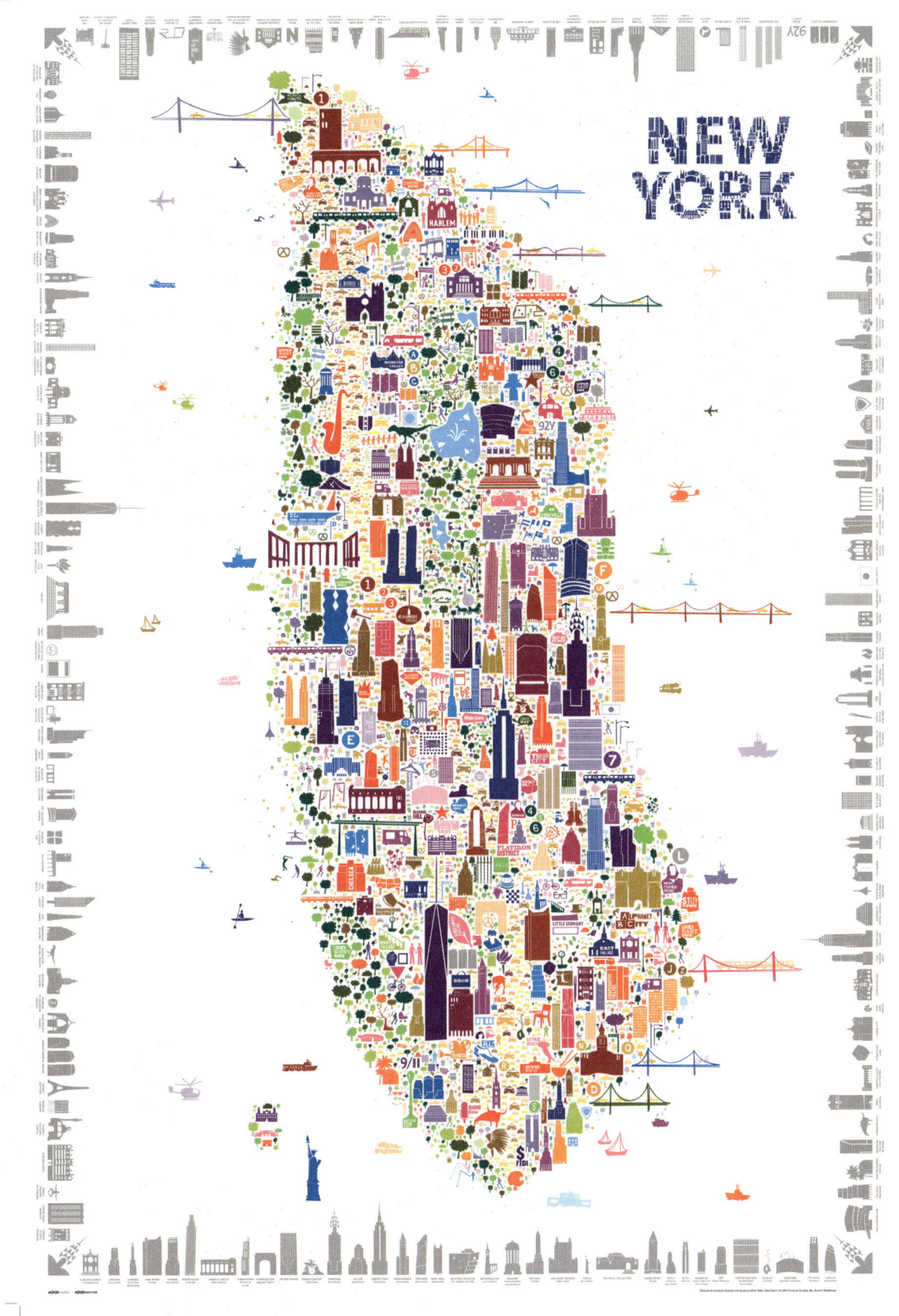

Iconic New York Poster

Rafael Esquer, the art director of this impressive piece, couldn't find the perfect New York souvenir for his friends and family when they visited. So he created it. Esquer has designed a museum-quality poster that is now becoming a favorite New York keepsake of visitors and residents alike. Painstakingly researched, drawn, and composed, Iconic New York™ spent two and a half years on the drawing board at Alfalfa New York. Esquer and his design team chipped away at the map of Manhattan, methodically converting the city's most beloved architecture and artifacts into artful graphic icons. Slowly but surely, from Washington Heights to Battery Park, they meticulously assembled the icons into a collage forming the shape of Manhattan itself. Iconic New York™ is printed on premium Italian heavy cardstock.

DESIGN Alfalfa Studio   COUNTRY/REGION The USA

MAPPING THE PHYSICAL ENVIRONMENT

Geografia-Twistable Globe

The globe packs two faces of the world into one cube, so it presents how the planet looks like from the space and how the space looks like from the earth. By twisting the looped eight cubes, you can convert a globe to a celestial globe.

DESIGN Drill Design   PHOTOGRAPHY Ryokan Abe   COUNTRY/REGION Japan

MAPPING THE PHYSICAL ENVIRONMENT

Geografia-Foldable Globe

A compact globe can be folded and carried. It can be carried in a small case or attached to a bag, so that you can look at it whenever you like.

DESIGN Drill Design   PHOTOGRAPHY Ryokan Abe   COUNTRY/REGION Japan

063

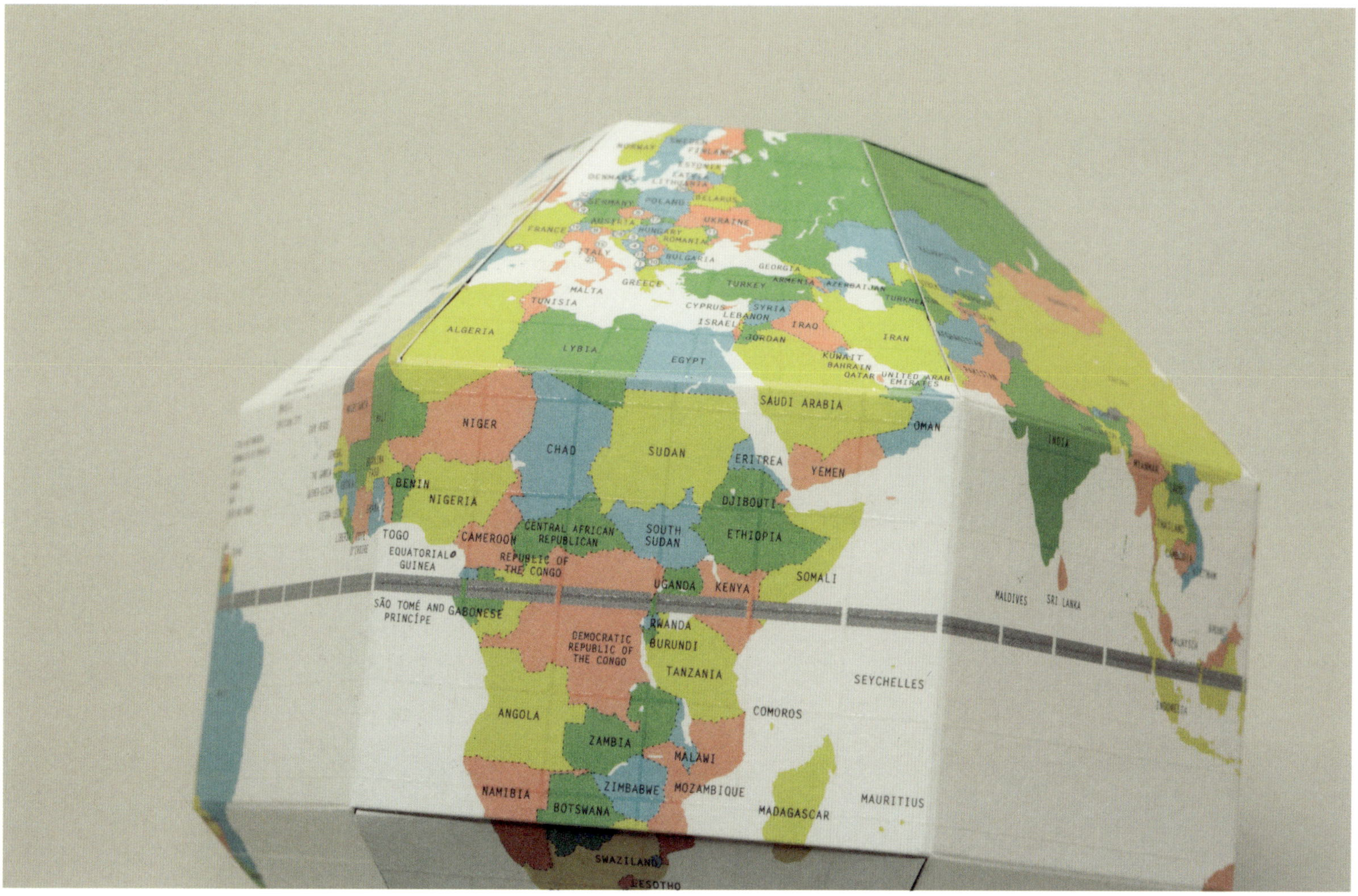

Summer in London Map

This canvas map was based on the theme of 'Summer in London'. It shows all the events and outside spaces in London such as Wimbledon, Gay Pride, and Notting Hill Carnival. Smaller but interesting things that London's citizens experience during the summer such as sweating on the subway, wasps and drinking outside can also be found on the map.

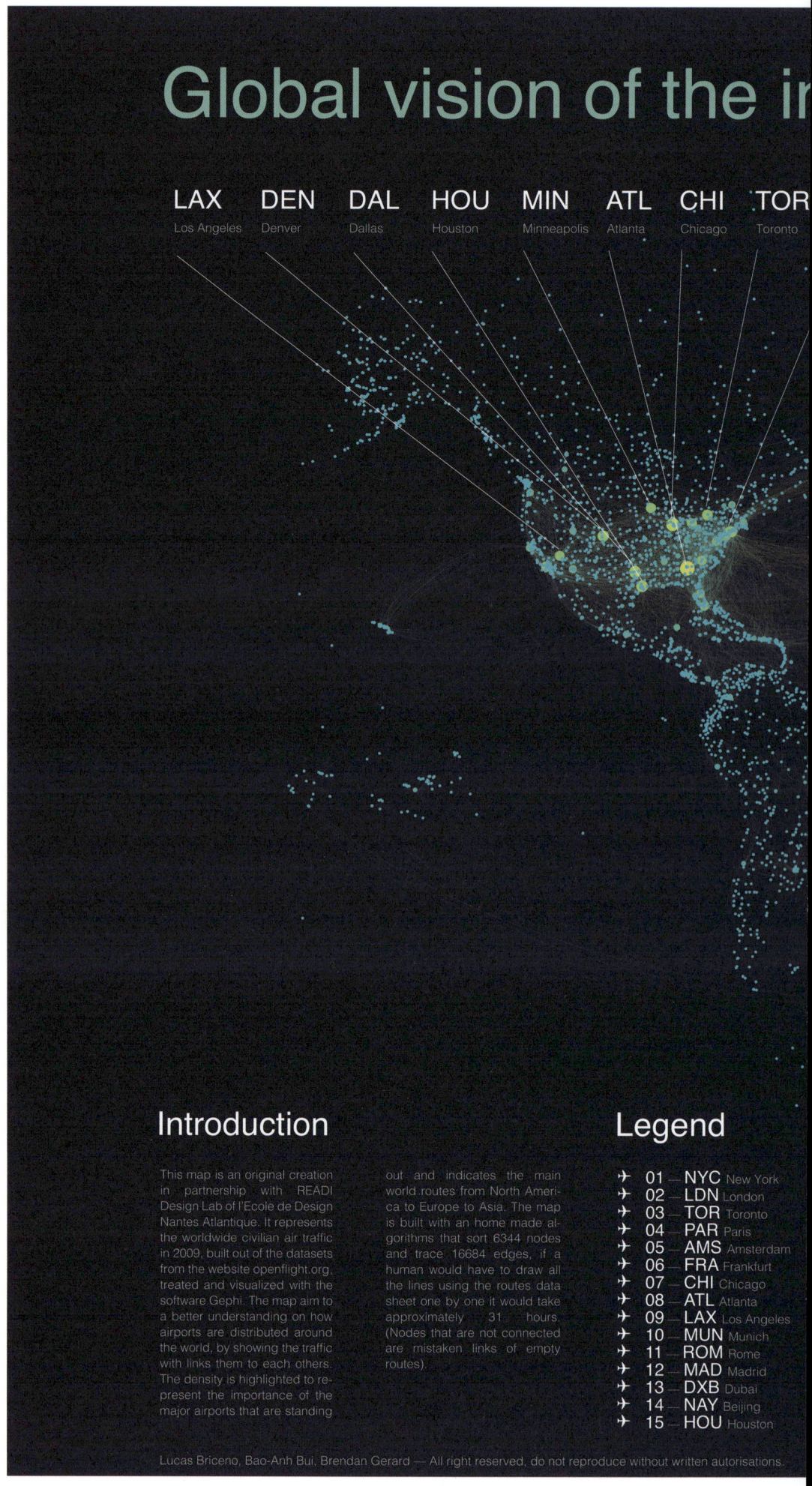

Flightviz

This map represents the worldwide civilian air traffic in 2009, built out of the datasets from the website openflight.org, treated and visualized with the software Gephi. It aims to visualize a better understanding of how airports are distributed around the world, by showing the traffic which links them to each other. The density is highlighted to represent the importance of the major airports and indicates the main world routes from North America to Europe and Asia.

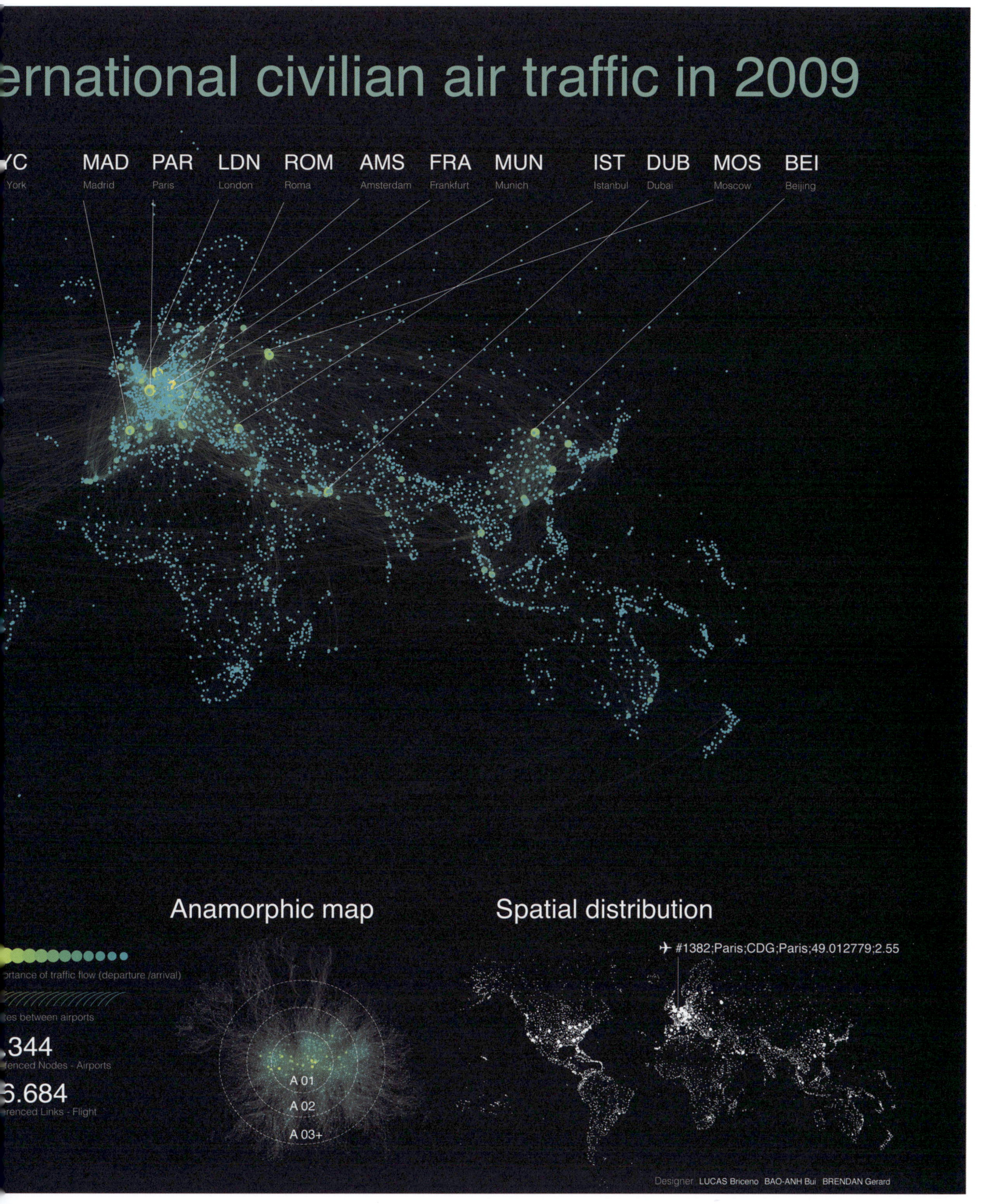

Los Angeles #1 Chinatown

Map Sculptures

These works map terrain of cities as well as their historical and cultural landscape, reflecting each city's evolving individuality and idiosyncrasy. The sculptures interweave the narratives of personal and public history. Within the urban form this builds a portrait of a city layered with history and time. Picton is less concerned with creating a factual and objective record than presenting an emotional and cultural history of the city, the non-objective mirror of history through the film, music, literature and visual art of the time.

Los Angeles #1 Chinatown is cut through on the surface from the poster "Chinatown". This film by Roman Polanski, starring Jack Nicholson and Faye Dunaway is about the corruption that accompanied the acquisition of the water rights from the Owens Valley in the early twentieth century. Partially visible underneath the sculpture is the poster from the film "LA Confidential" a film that also illuminated the legendary corruption of the Los Angeles Police Department in the 1950s.

Los Angeles #2 Bladerunner

Los Angeles #2 Bladerunner is an extensively fictionalized city, where the line between actual history and that of a screen play are so completely blurred and intertwined.
This sculpture is created from a cut through poster of the classic neo noir film "Bladerunner." Behind this are fragmented book covers from some of the classics of Noir fiction to have emerged from Los Angeles, such as The Big Nowhere, White Jazz, LA Confidential and The Black Dahlia, by James Elroy, and Mildred Pierce and Double Indemnity by James McCain, etc. In between these layers and following the main street are lines of text and quotes relating to the history of Los Angeles. Finally the quote which runs the circumference of the work is by Carey Mc Williams from the 1940s: "The belief in some awful fate that will someday engulf the region is widespread and persistent."

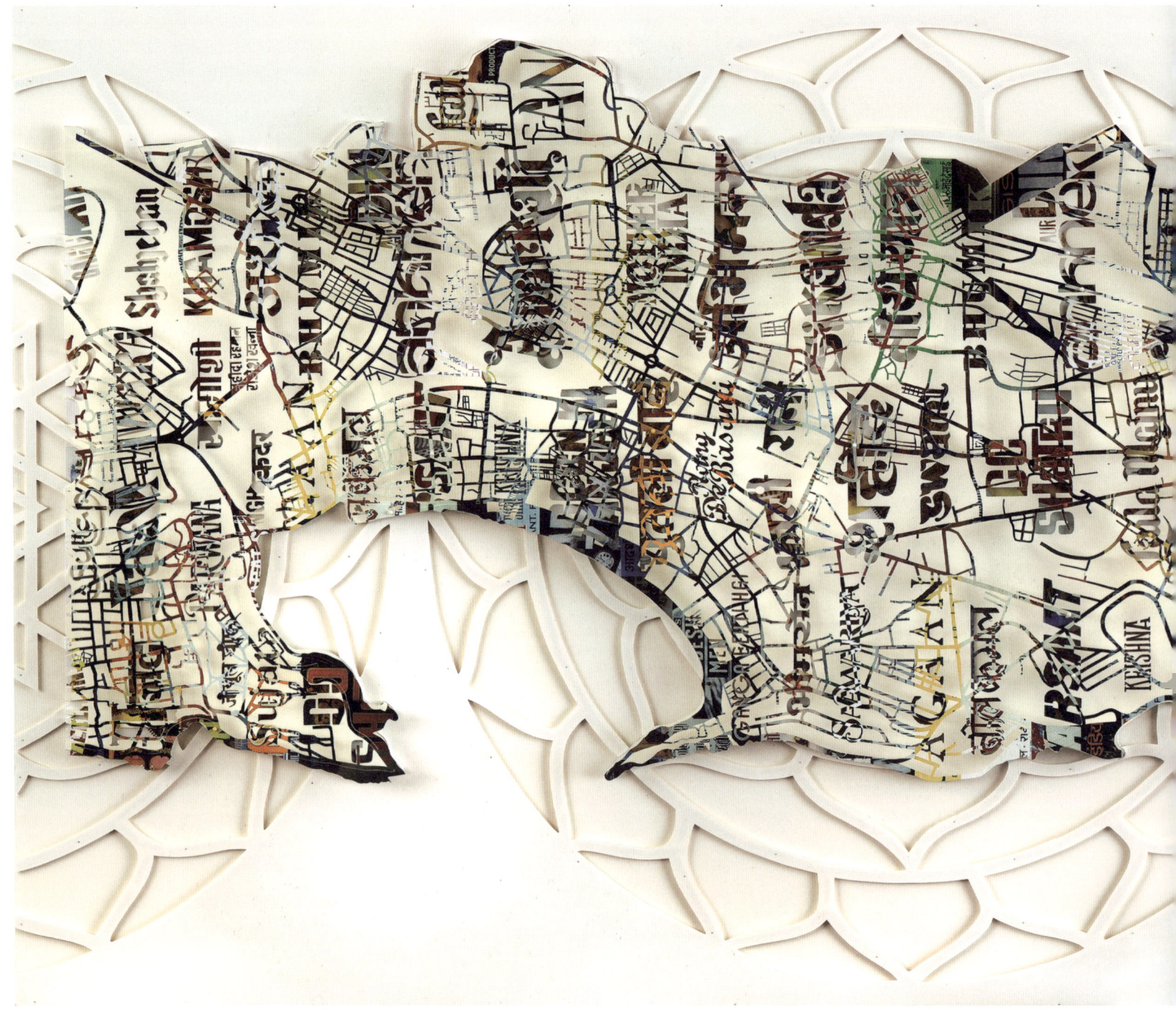

Mumbai/ Bombay

The street maps of Mumbai have been created out of Bollywood film posters.
Surrounding the form of the peninsula of Bombay in the ocean are white cut out forms of the Shri Yantra, representing the beauty of three worlds, earth, atmosphere and sky, and the union of masculine and feminine divinity. This is one of the forms created for the annual Diwali festival in Mumbai. Diwali is a celebration of inner light emerging from spiritual darkness, it is an ancient Hindu festival that takes place every autumn and centers on the new moon. The night of Diwali is the day Lakshmi chooses her husband Vishnu as her husband and marries him. Prayers are typically offered to Lakshmi, goddess of prosperity; lamps are lit to symbolize the sun, cosmic giver of light and energy to all life. Outside the houses in the doorways and entrances to their homes, women and children create Rangoli patterns in bright colors to surround their homes with auspicious symbols for the year ahead.
The entire sculpture is underpinned and surrounded by the Shri Yantra, which is perhaps the most meaningful symbol of all.

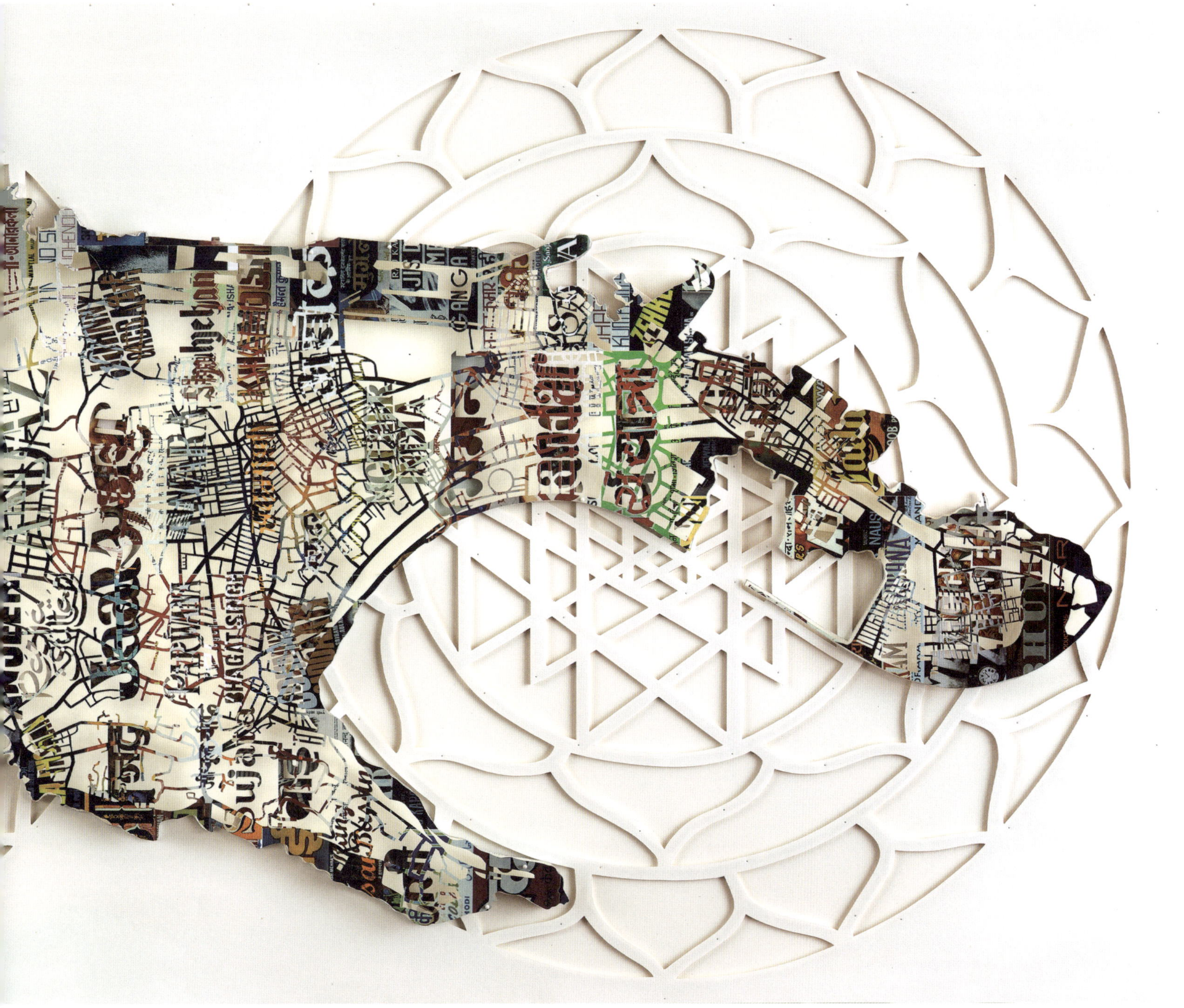

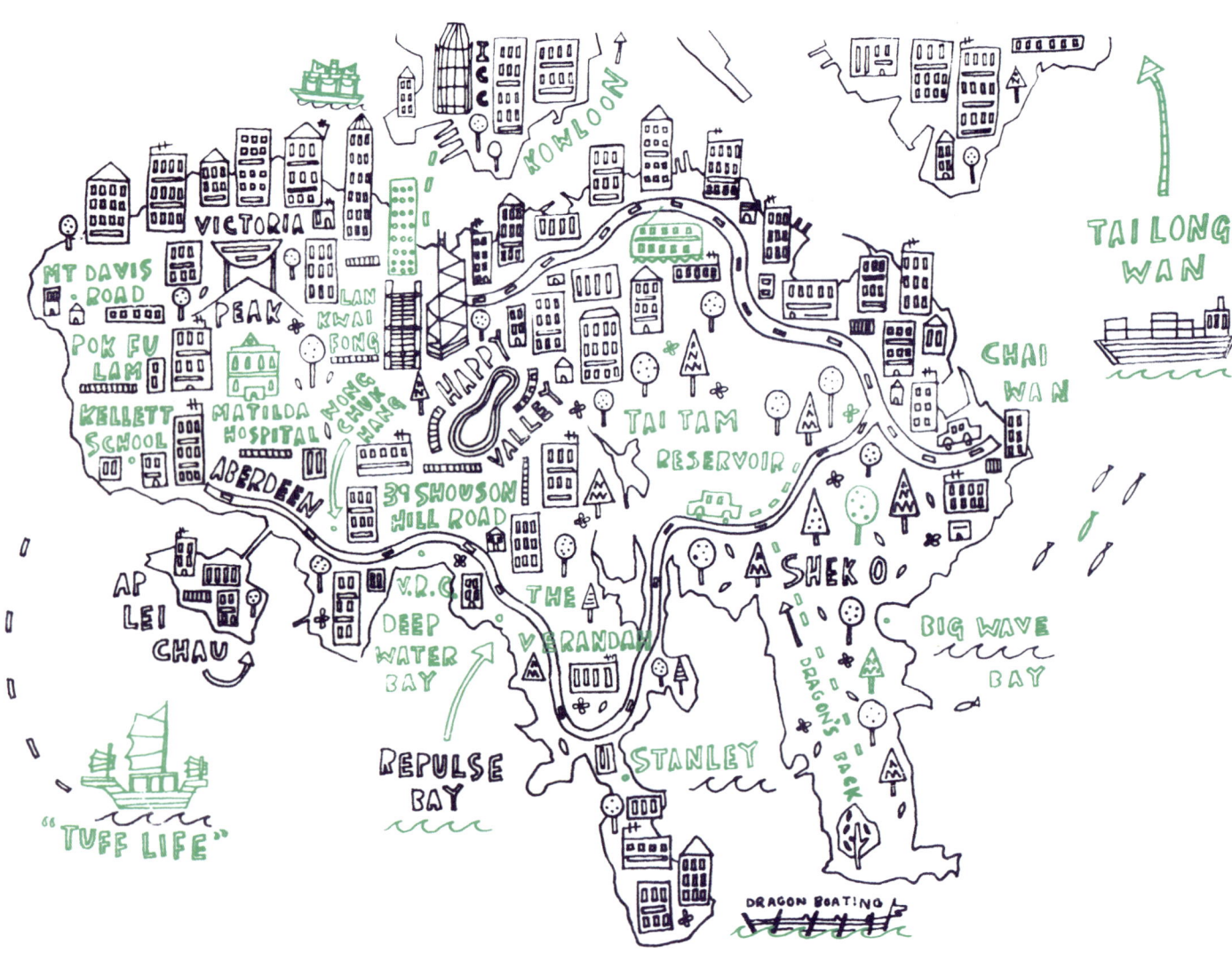

Hong Kong Illustrations

Both the illustrations are from Marshall's city map series. The first one features the whole of Hong Kong island. Marshall designed the map based purely on line, incorporating well known landmarks with places of personal significance to the client. The illustration was designed as a two-color screen print with key features highlighted in a contrasting color.

Similarly, the one on the right used contrasting colors with key features highlighted in red and other elements in blue. Based on the client's demands, the map incorporated key landmarks, including the Bank of China building and I.C.C. alongside places of more personal significance.

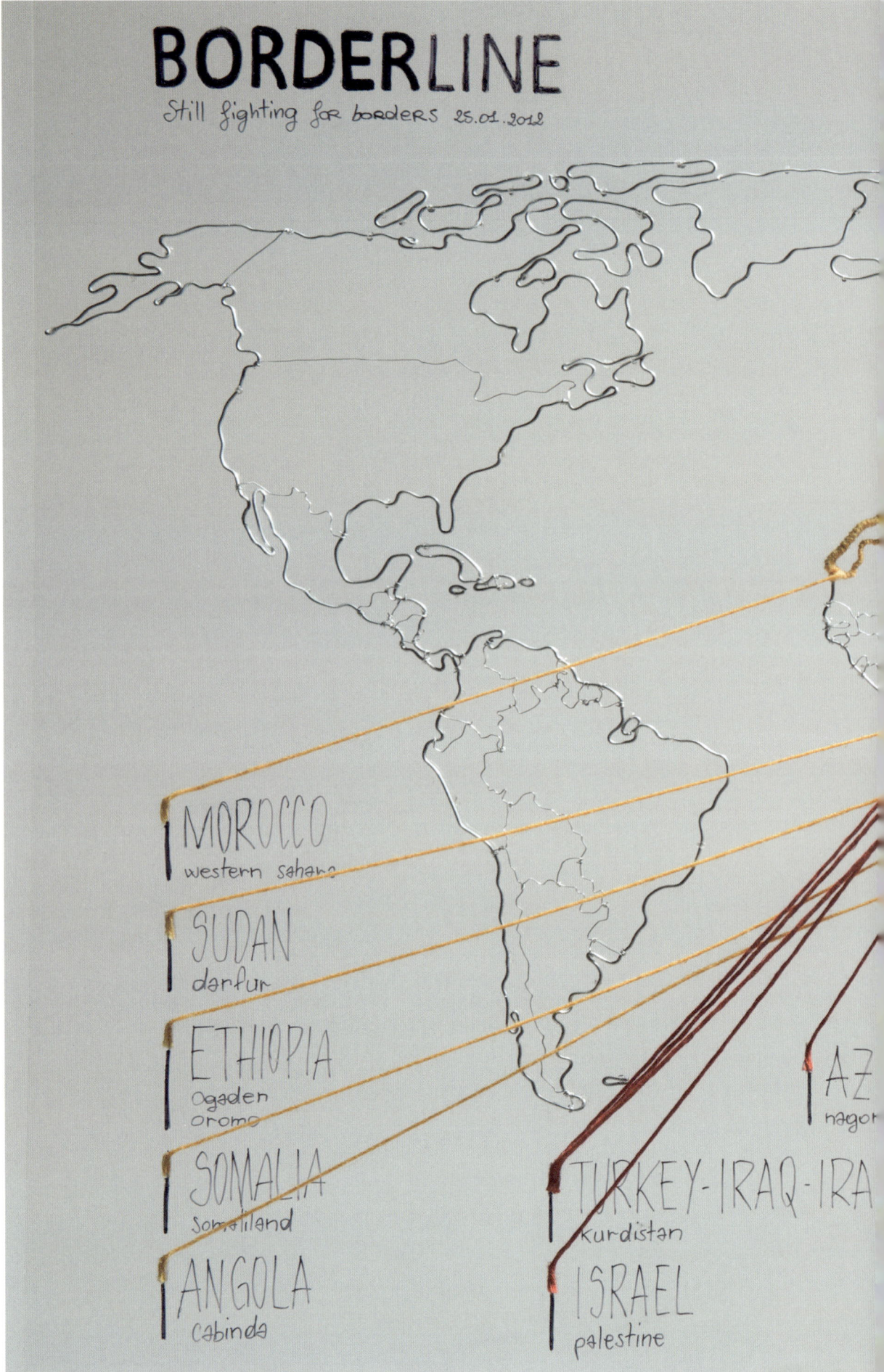

Borderline

The aim of this project is to reflect upon the purpose of national borders.
To illustrate this idea, the map was made out of flexible metal wires on a transparent board. A few of the borders have been highlighted with colored threads to express the possibility of upcoming alteration due to internal turmoil or fight for independence. The project gives us an opportunity to think about the violence caused by something that is, by its very nature, arbitrary and unstable.

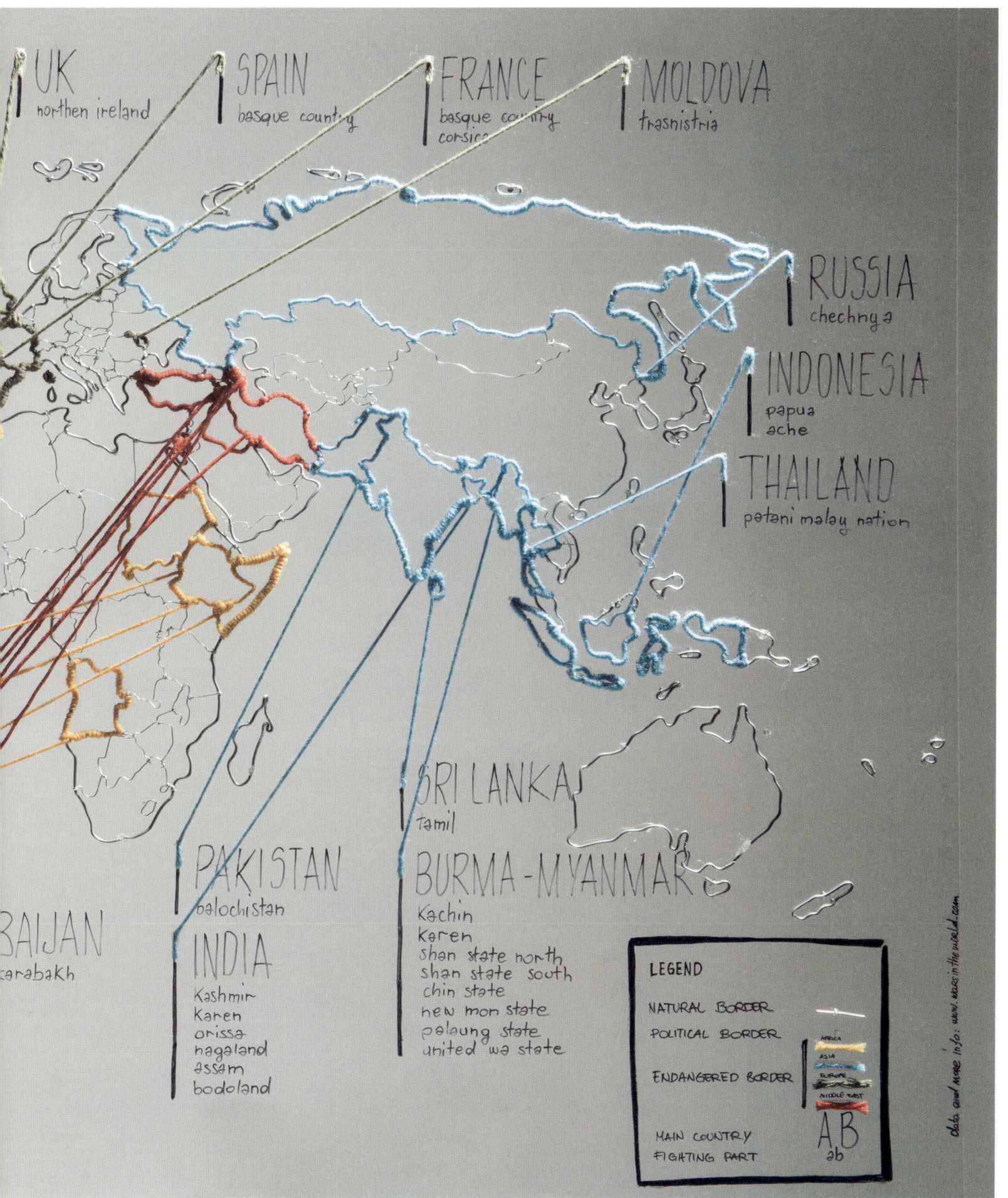

MAPPING THE PHYSICAL ENVIRONMENT

"Around the World" Illustrated Map

With its colorful design, the Around the World map takes you on a journey of discovery. Featuring famous landmarks, the map can be used as an educational tool for children or to feed an exploratory mind. The map is printed on a magnetic board where kids can mark with a dry erase marker or place magnets on special places they have been or would like to travel to.

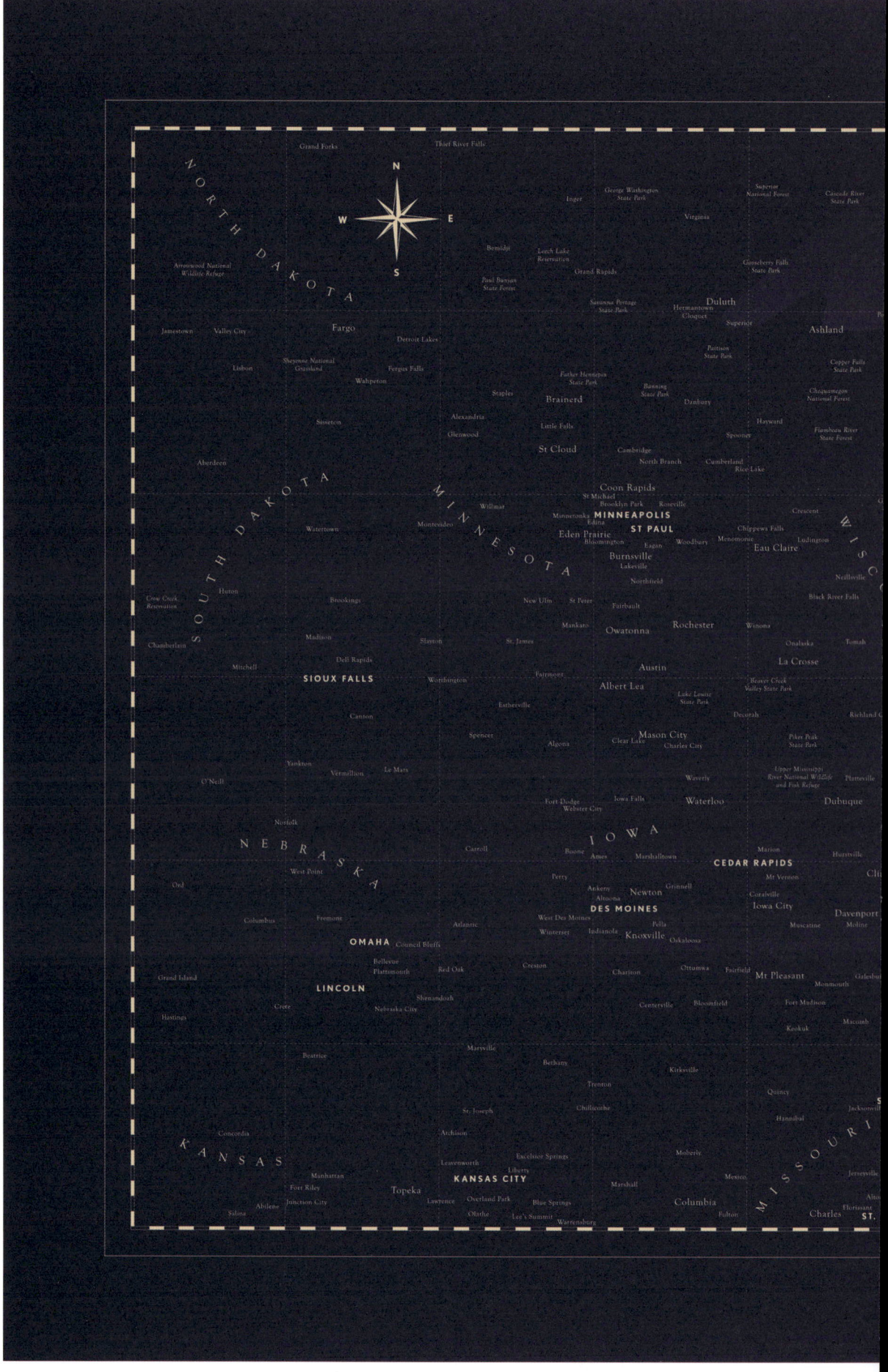

Typographic Heartland Map

This is a map of the Midwest of the United States. The aim was to create the map out of purely typography using funtional information. Though this point is still abstract in the resulting map, it results in a cool visual effect. The map was printed on Patriot Blue Classic Crest 80lb Cover, with a cream colored ink. The Great Lakes were hit with a gloss varnish to add another subtle dimension to the map.

Bath °c Thermo Colour Map

This "magic" map is a hand held activated map, based on Bath city, England. Using thermo chromic inks and tyvek fabric the map is activated at different temperatures, revealing layers of hand illustrated buildings and attractions, showing the best places to visit depending on the weather. Designed as a roll map, the fabric is waterproof and crumpleproof, allowing it to be easily stored in a bag. The map is color coded to specific external environmental temperatures, which allows tourists, visitors and residents alike to have a new experience of Bath.

DESIGN Camilla Hempleman-Adams COUNTRY/REGION The UK 081

Map Art

Tompsett likes to experiment with different styles and subjects, but his main focus is on map art and city skylines where he has become a trend setter. These three artworks are picked up from his map art series. The World Map Painted Splashes is a map created from gold, burgundy and purple paint splashes. And the other two maps depict the city of San Francisco in different color palettes. The color-rich one draws land areas filled with five selected colors and the sea in light blue, while the other one is a watercolor background with the land areas overlaid and darkened.

SAN FRANCISCO

Mysteries and Curiosities of the Jet Propulsion Laboratory

NASA Jet Propulsion Laboratory is a place where employees need to use GPS to find their way around the Lab. For one, buildings have numbers instead of names. Second, buildings are identified by the number of when they were funded, instead of by location. For example, Building 67 is perplexingly located between Buildings 238 and 138. So the design team plotted this map which could be a suggested guide (but not an authoritative one) for tourists to this mysterious and curious place known as the Jet Propulsion Laboratory. The map itself is divided into two sections. The front is an Insider's Guide to JPL. The back provides several Walking Tours. These tours encourage Lab exploration. Whether visiting the world's most stable clock, the weavers who hand-stitch the thermal blankets for every mission, or simply finding a new place to have lunch, this map offers a fresh perspective on the overlooked aspects of the Lab's culture that make it so unique.

Here's Your Jetpack
AN INSIDER'S GUIDE TO JPL

A Brief History of Time

October 31, 1936 — The Suicide Squad performs its first motor tests in the Arroyo Seco near Devil's Den.

July 1, 1944 — The Lab is officially named the Jet Propulsion Laboratory.

January 31, 1958 — Explorer 1 becomes the first satellite launched by the United States.

January 1, 1959 — JPL transfers from the U.S. Army to NASA.

August 27, 1962 — Mariner 2 becomes the first spacecraft to successfully fly by another planet and return data.

November 28, 1964 — Mariner 4 returns the first images ever taken of another planet's surface by a spacecraft.

June 2, 1966 — Surveyor 1 makes the first American soft-landing on the Moon.

March 18, 1966 — Goldstone's 64-m antenna receives first spacecraft signal. Goldstone is one of three structures that would later become NASA's Deep Space Network.

JPL Through the Looking Glass

PEOPLE, OUR GREATEST ASSET
JPL is a community of trailblazers, constantly pushing the limits of space exploration. United by a common mission, we seek to go where no one has gone before.

Groups   Sections   Divisions

THE BUSINESS OF SPACE TRAVEL
Every business has an organizational structure and JPL is no different. Each JPL employee is assigned to a Group. This Group, along with other Groups, makes up a Section. Several Sections constitute a Division, and several Divisions form a Directorate.

DIRECTORATES
The greater JPL pie is divided into nine slices called Directorates. Each Directorate is charted with its own role and responsibilities ranging from business administration and safety, to research into Earth, Mars, and Deep Space.

Rocket Men

Meet the Directors who have transformed JPL from a rocket-building facility to one that conducts space exploration.

THEODORE VON KÁRMÁN — 1938–1944 — Dr. von Kármán made considerable contributions to the field of aerodynamics, and was awarded the first National Medal of Science by President Kennedy.

FRANK MALINA — 1944–1946 — Dr. Malina helped establish the research and testing facilities that would become JPL. He also helped develop many of JPL's early successes for the U.S. Army.

LOUIS DUNN — 1946–1954 — Under Dr. Dunn's leadership, JPL developed its first post-war guided missile, the Corporal.

WILLIAM PICKERING — 1954–1976 — Dr. Pickering led JPL's efforts to launch the first U.S. satellite, Explorer 1. He also won JPL's transfer from the U.S. Army to NASA.

BRUCE MURRAY — 1976–1982 — Dr. Murray was part of the imaging team for the early Mariner missions to Mars and was a forceful advocate for planetary exploration. After his JPL tenure, he returned to Caltech, where he is currently an emeritus professor.

LEW ALLEN, JR. — 1982–1990 — Dr. Allen had a long and successful career in the military before becoming JPL's Director.

EDWARD STONE — 1991–2001 — With a specialty in cosmic radiation, Dr. Stone became project scientist of the Voyager mission. After a decade as Director, Dr. Stone returned to Caltech where he is still a Voyager project scientist.

CHARLES ELACHI — 2001– — With an expertise in imaging radar, Dr. Elachi helped develop instruments that study Venus and Earth. He is currently the science team lead for the Cassini spacecraft's imaging radar instrument.

Source: www.jpl.nasa.gov

Eureka!

JPL has more patented intellectual property than all the other NASA centers combined. In fact, more than 200 U.S. companies have taken advantage of JPL's innovations. The inventors on Lab have contributed to commercial use in the fields of:

- Charge-coupled detectors or CCDs (now widely used in digital cameras)
- Advanced GPS receivers
- Image-processing techniques such as image restoration, mosaic effects, map projects, and three-dimensional "fly-through" animations
- Wireless communications, including deep space antennas, precision-timing systems, signal detection, and digital processing
- Digital imaging such as CAT scanning, ultrasounds, cardiac angiography, and nuclear magnetic resonance
- Miniaturization such as miniature infrared sensors used to locate cancerous tumors, and tiny camera chips placed in pills to photograph the digestive system
- Robotics, including robotic microsurgery, autonomous operations, and hazard avoidance visualization
- Lightweight optics and wavefront sensing and control
- Electronic monitoring systems that can remotely detect concealed weapons, heartbeats, and biometrics

Sources: JPL 101; "Into the Black" by Peter J. Westwick; National Space Technology Applications Office (NSTA).

The Suicide Squad

JPL was started by a group of Caltech students, a high school dropout, and a known occultist. Dubbed the "Suicide Squad", the group earned its nickname after repeatedly escaping explosions and asphyxiation during their early rocket tests. Like any team of super heroes, each member brought a unique talent to the group.

APOLLO M.O. SMITH — pipelines & inflammables specialist
not pictured: TSIEN HSUE-SHEN — mathematician
WELD ARNOLD — money man
RUDOLPH SCHOTT* — student assistant who "wanted to watch"
FRANK MALINA — theoretician
ED FORMAN — mechanic
JACK PARSONS — explosives expert

*Not a member of the Squad

Lab Lingo

As when joining any exclusive club, new JPL hires need to know certain lingo. Here is a list of essential terms and definitions to help ease your transition.

AEROGEL: The lightest solid ever manufactured at 99.8 percent air. Also referred to as "Solid Smoke."

CASH OR CARD: The cafeteria does not accept credit or debit cards. Rather, it only accepts cash or a Caltech Dining Card. Thank you, darling.

CLEANROOM: A specially constructed area that provides a controlled environment for chemical, microbial, radiological, and/or particle contamination control.

GRAVITY ASSIST: A clever method of using energy from a planet to help speed a spacecraft on its way.

JPLer: A JPL employee.

LANYARD: The material used to hold up an employee badge — the most coveted being a mission lanyard.

PI: A Principal Investigator (PI), the science lead on a mission.

PURPLE PIGEON: An outside-the-box idea that could lead to a future mission.

RDO FRIDAYS: A regular day off, or RDO, that allows employees who work a 9/80 schedule (i.e., 80 hours of work in 9 days) to take every other Friday off from work.

SYSTEMS ENGINEERING: An interdisciplinary field that focuses on how complex engineering projects should be designed and managed over the life cycle of a project. JPL is a leader in this field.

TIGER TEAM: A small, nimble group of a mission's best people who are pulled together to quickly diagnose and fix a problem.

QUIET HOUR: Not a nap, but rather a meeting with your manager where you can talk about anything.

WAM: A work authorization memo (WAM), the process that enables JPL managers/supervisors to delegate and authorize work for a specific cost account (i.e., the way you get paid).

Safety Signs A–Z

Because of the dangers inherent in rocket science, JPL has taken extra steps to ensure that safety is a top priority. Below is a sample of the large variety of safety signs on lab that you may not see at another place of business.

        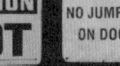 

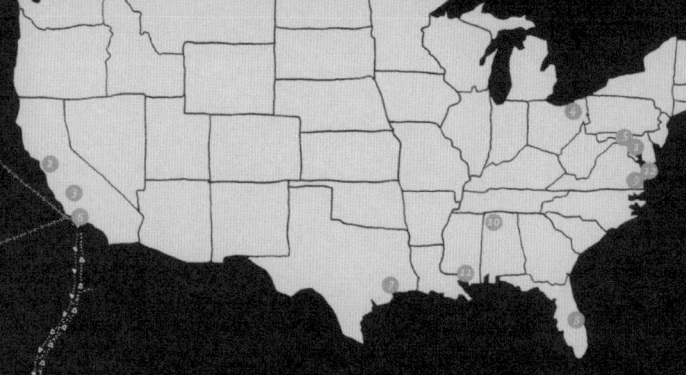

...ber 14, 1971
...9 becomes
...st spacecraft
...t Mars.

August 20 & September 5, 1977
Voyager 1 and 2 fly by Jupiter and Saturn while Voyager 2 also visits Uranus and Neptune. The Voyager missions would become JPL's longest-lived, with both spacecraft currently operating and returning data.

June 28, 1978
Seasat, an experimental satellite, flight-tests four instruments that use radar to study the Earth and its seas. Many subsequent Earth-orbiting instruments developed at JPL owe their legacy to this mission.

May 4, 1989
Magellan launches and later uses imaging radar to map 99 percent of the Venetian surface over four years.

December 2, 1993
An optical flaw is discovered in the Hubble Space Telescope's main mirror. JPL's Wide Field and Planetary Camera 2 saves the mission by correcting the space telescope's vision.

February 12, 2000
The Shuttle Radar Topography Mission acquires enough data to obtain the most complete near-global mapping of our planet's topography to date. Google uses these images for the first version of Google Maps.

January 4 & 25, 2004
Two mobile geologists named Spirit and Opportunity land on Mars. The rovers successfully complete a three-month primary mission on opposite sides of Mars and continue an extended mission for over seven years.

Source: www.jpl.nasa.gov

November 4, 2010
EPOXI encounters Comet Hartley 2. The EPOXI mission reused the Deep Impact spacecraft bus to catch up to another comet. This is the first spacecraft to visit two comets!

An Insider's Guide to the Mysteries & Curiosities of the Jet Propulsion Laboratory

177 Acres / +150 Buildings on Lab / ~5,000 Employees / ~1,000 PhDs / 2 National Historic Landmarks / TV Studio, Mars Yard, and Extraterrestrial Materials Lab / Expertise in Planetary Exploration, Earth Science, Telecommunications, Astronomy, and Physics / Developed and Manages the Deep Space Network / Lead U.S. Center for Robotic Exploration / Missions to Every Planet in the Solar System

NASA Centers

JPL is a division of Caltech but is also a federally funded research and development center (FFRDC). JPL is the only NASA Center that is managed by a university.

1. (HQ) NASA Headquarters: Washington, DC
2. (ARC) Ames Research Center: Moffett Field, CA
3. (DFRC) Dryden Flight Research Center: Edwards, CA
4. (GRC) Glenn Research Center: Cleveland, OH
5. (GSFC) Goddard Space Flight Center: Greenbelt, MD
6. (JPL) Jet Propulsion Laboratory: Pasadena, CA
7. (JSC) Johnson Space Center: Houston, TX
8. (KSC) Kennedy Space Center: Cape Canaveral, FL
9. (LARC) Langley Research Center: Hampton, VA
10. (MSFC) Marshall Space Flight Center: Huntsville, AL
11. (SSC) Stennis Space Center: Bay St. Louis, MS
12. (WFF) Wallops Flight Facility: Wallops Island, VA

Office of the Director
Business Operations Directorate
Engineering and Science Directorate
Solar System Exploration Directorate
Office of Safety and Mission Success
Mars Exploration Directorate
Astronomy, Physics, & Space Technology Directorate
Earth Science & Technology Directorate
Interplanetary Network Directorate

Signs of Life

"They may not wear lab coats or design spacecraft, but the wildlife that shares JPL's 177-acre campus is also part of its culture along with the human inhabitants. The Lab was built from the arroyo, or dry creek, up into the side of the San Gabriel Mountains. Lab employees learn quickly that they are simply allowed the courtesy of sharing the land." — JPL 101

NOCTURNALS
Coyote (3)
Opossum
Gray Fox
Mountain Lion

BIRDS
Red-tailed Hawk (4)
Great Horned Owl
Blue Jay
Peregrine Falcon
Least Bell's Vireo
Coastal California Gnatcatcher

DIURNALS
Mule Deer (1)
Bobcat
Ring-tailed Cat
Tree Squirrel
Ground Squirrel

REPTILES
Garter Snake (2)
Gopher Snake
King Snake
Rattlesnake
Arroyo Southwestern Toad

TREES
Australian Fire Wheel (5)
Olive Tree (6)
Citronella Lemon-Scented Eucalyptus (7)
Coral Tree (8)
Brazilian Pepper Tree
Dawn Redwood Tree
Silk Floss Tree
Manzanita

INSECTS
Black Widow
Tortoise Beetle

PLANTS
Poison Sumac

Tracks clockwise from top left: Coyote, Bobcat, Ring-tailed Cat, Ground Squirrel, Mule Deer, Opossum

The Caltech Connection

Who does a JPL employee work for—Caltech or NASA? JPLers are employees of Caltech, not civil servants. In contrast, the buildings in which they work, all the way down to the furniture and pencils, are property of the U.S. Government.

JPL is approximately 7 miles or a 16 minute drive (without traffic) to the California Institute of Technology. This is roughly the route you would take when traveling between the two institutions.

Source: Google Maps

Cosmic Callings

The odd cadre of jobs on Lab shows that many JPLers did not follow the advice of their high school guidance counselors. It is this wide range of skills, working toward a common mission, that make space travel possible. Here is a list of JPL's Cosmic Callings:

Auto Mechanic
Ethics Advisor
Firefighter
Glassblower*
Historian
Industrial Hygienist
Interplanetary Navigator
Locksmith
Magnetic Cleanliness Engineer
Mail Carrier
Microdevices Engineer
Parking Program Coordinator
Risk Communication Specialist
Thermal Blanket Weaver
Traffic Analyst
Visual Strategist

*Retired

Introduction

For a place that depends on logic and reason, the Lab's layout is anything but. In fact, a running joke at JPL is that its employees need to use GPS to find their way around the Lab. For one, buildings have numbers instead of names. Secondly, buildings are ordered in the number in which they were funded, instead of by location. For example, Building 67 is located between Buildings 238 and 138.

Intrigued by this dichotomy and wanting to know more about JPL aside from the four walls of my cubicle, I came up with a plan. Armed with a GPS tracking device, camera, and a trusty pair of shoes, I walked to every building on Lab in numerical order. What I thought would take a Saturday afternoon took 22 hours over the span of four days at a walking distance of 52.2 miles.

The resulting map is a reflection of this wacky experiment, research at the Lab's library, and conversations with other JPL employees. The map itself is divided into two sections. The front is an Insider's Guide to JPL containing information I wish someone had explained to me when I began working at the Lab.

The back provides several Walking Tours. In the same way that JPL encourages space exploration, these tours encourage Lab exploration. Whether visiting the world's most stable clock, the weavers who hand-stitch the thermal blankets for every mission, or simply finding a new place to have lunch, this map offers a fresh perspective on the overlooked aspects of the Lab's culture that make it so unique.

The map is not meant to be an authoritative guide, but rather a conversation starter, a poster, or simply a celebration of the mysterious and curious place known as the Jet Propulsion Laboratory.

— Luke Johnson

Website: mysteries.jpl.nasa.gov

Questions & Comments: mysteries@jpl.nasa.gov

CURATION
Luke Johnson

DESIGN
Erin Ellis, Christiane Holzheid

COLLABORATORS
Dan Goods, Nora Mainland, Anisse Gross, Catherine Haradon

THANKS TO
Cozette Hart, Becky Campos, Victor Luo, Randii Wessen, Frank O'Donnell, Joe Courtney, Hunter Sebresos

FIRST EDITION

MAPPING THE PHYSICAL ENVIRONMENT

Pure Journeys

This detailed map allows users to explore New Zealand and shows what the country can offer to tourists and adventurers. All the islands are represented and the map showcases their quirks and treasures. Espinosa did research to capture each region's vibe and highlights. He also tried to achieve a balance between whimsicality and realism, because the map is meant to represent real driving routes around the country.

DESIGN Pablo Espinosa   COUNTRY/REGION Mexico

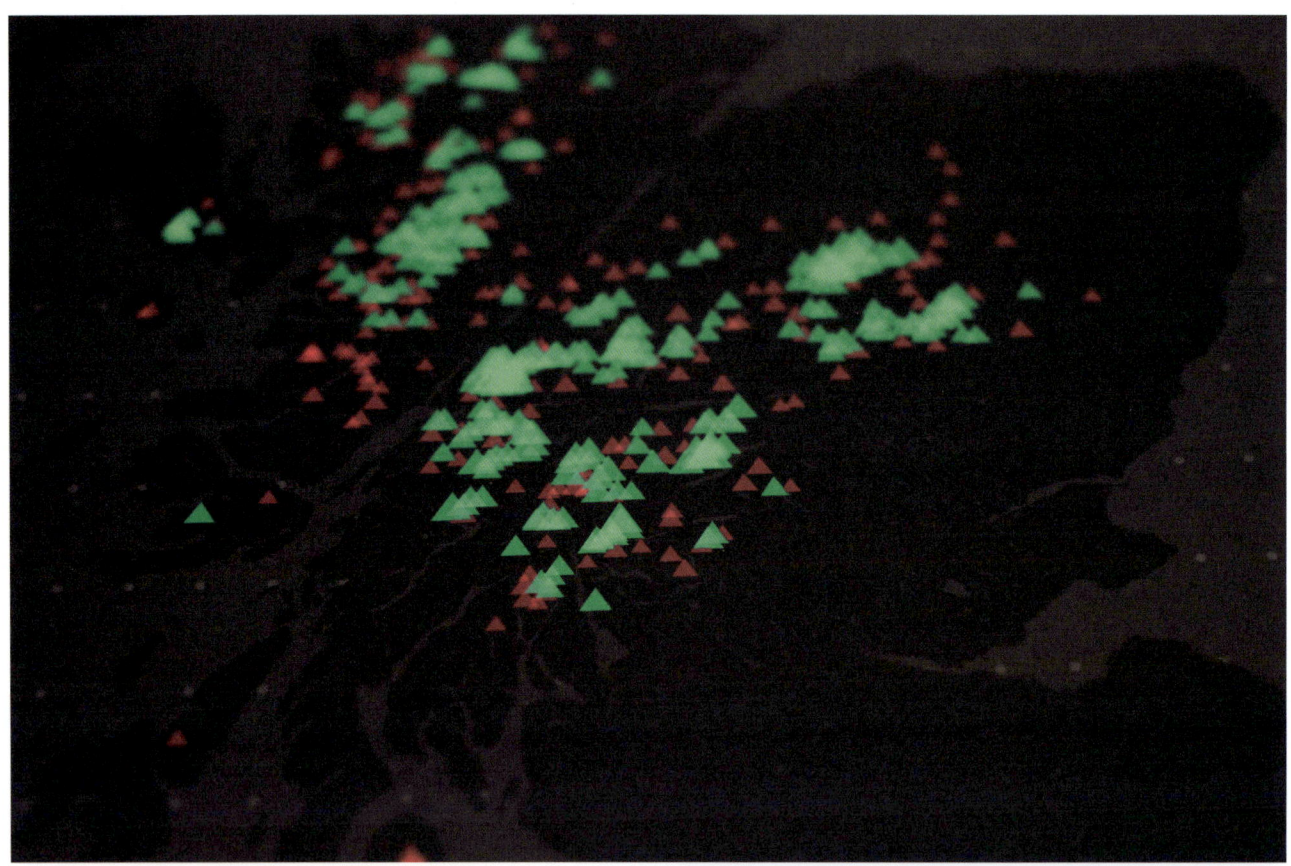

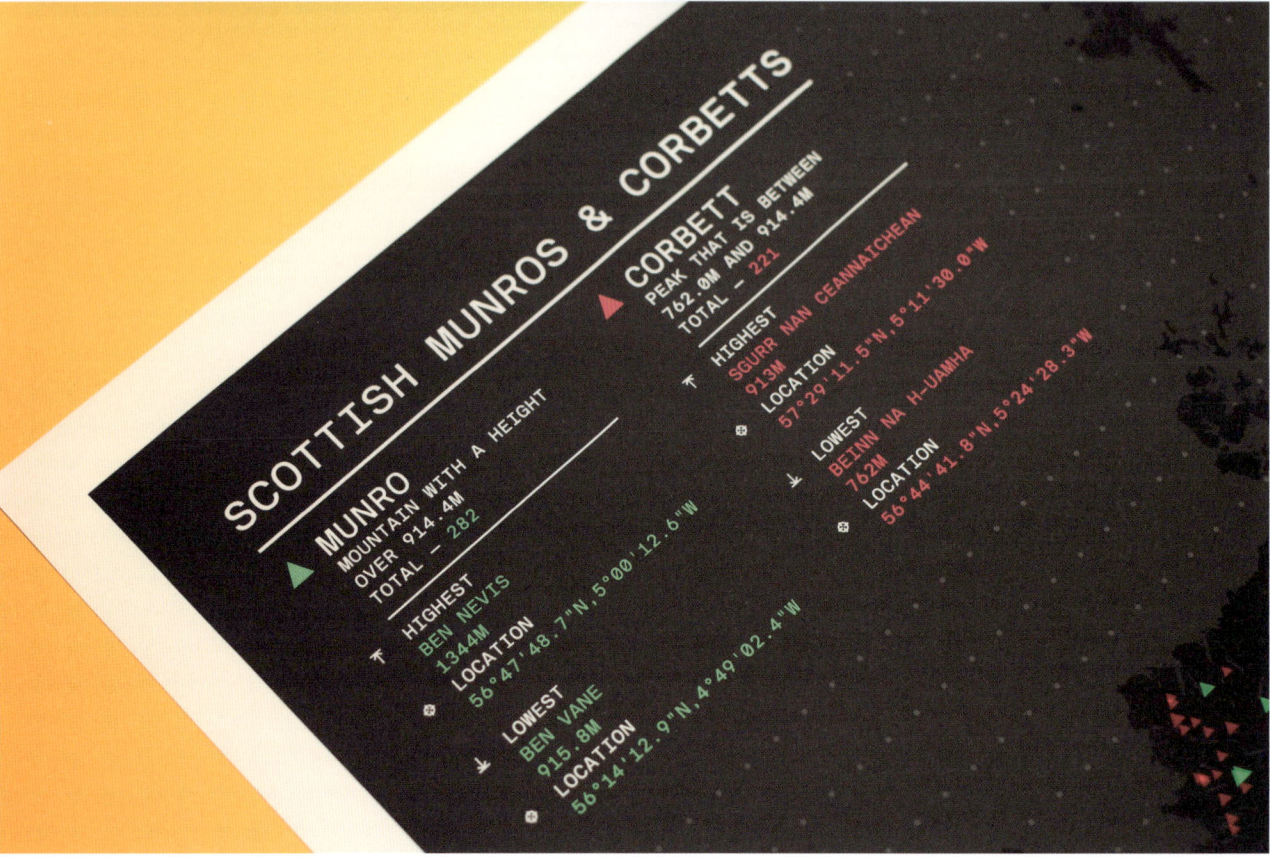

Processing: Scottish Mountains

Processing was used to produce this map of Scotland's Munros & Corbetts (types of Scottish Mountains). This technique allowed for markers to be precisely positioned and scaled by mountain height. The designer achieved both creative style and accuracy for this composition.

DESIGN Paul Mullen   COUNTRY/REGION The UK

SCOTTISH MUNROS & CORBETTS

▲ **MUNRO**
MOUNTAIN WITH A HEIGHT OVER 914.4M
TOTAL — 282

⛰ HIGHEST
BEN NEVIS
1344M
✦ LOCATION
56°47'48.7"N,5°00'12.6"W

⬇ LOWEST
BEN VANE
915.8M
✦ LOCATION
56°14'12.9"N,4°49'02.4"W

▲ **CORBETT**
PEAK THAT IS BETWEEN 762.0M AND 914.4M
TOTAL — 221

⛰ HIGHEST
SGURR NAN CEANNAICHEAN
913M
✦ LOCATION
57°29'11.5"N,5°11'30.0"W

⬇ LOWEST
BEINN NA H-UAMHA
762M
✦ LOCATION
56°44'41.8"N,5°24'28.3"W

Map of Mesoamerica

The map of Mesoamerica shows the main sites of the pre-classic Maya culture in its nearly original condition. This map is not only a topographic overview with architectural structures; it also gives an inside look to important eras and puts them into a new context. The Chicxulub impact in the north of Yucatán formed the landscape and laid the foundation of human kind, the ancient Maya and also of Pepsi-Cola for human civilizations. To capture the feel of this whole area, the map is colored like a landscape and lightning.

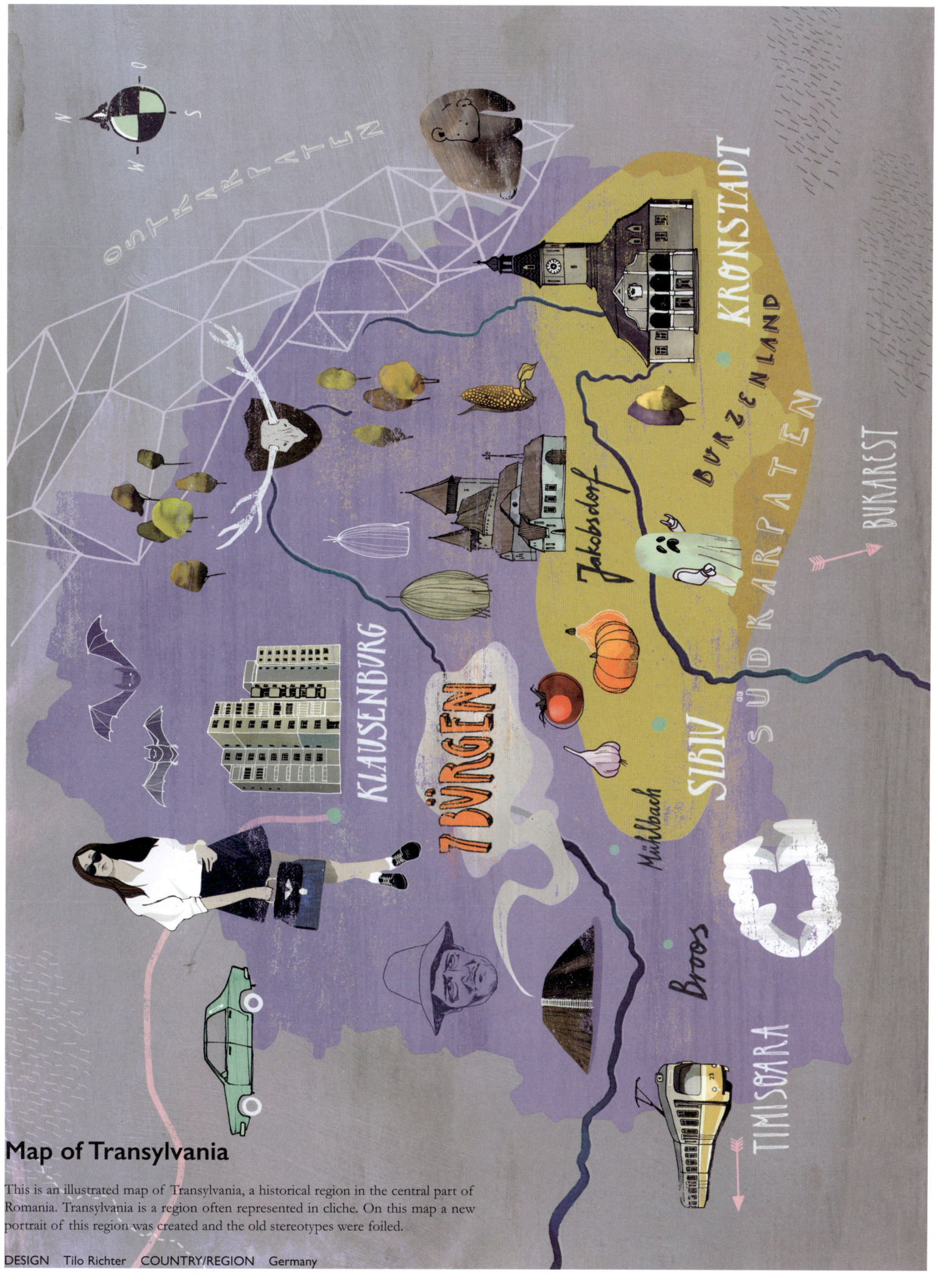

Map of Transylvania

This is an illustrated map of Transylvania, a historical region in the central part of Romania. Transylvania is a region often represented in cliche. On this map a new portrait of this region was created and the old stereotypes were foiled.

DESIGN Tilo Richter   COUNTRY/REGION Germany

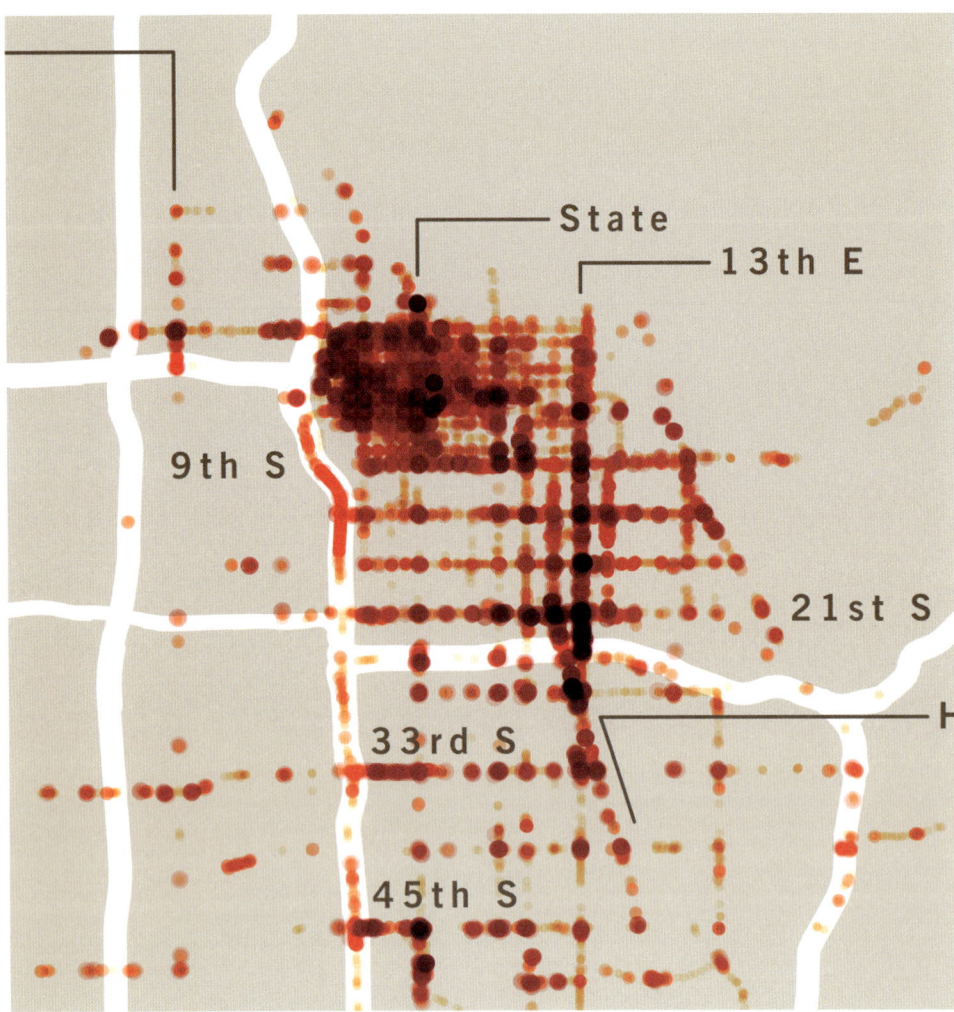

Stop Map

During five years of running around the Salt Lake City-Provo Metro, Hull tracked a lot of traffic and stop lights, filtering all those tracks for speed, the patterns of where the freeway backs up and which major roads have a better flow contrasting with ones that are commonly slow. In town, the designer depicts unsurprising high traffic areas in downtown Salt Lake, Sugarhouse, Park City, and State Street in Orem. On the interstate, few people are surprised by the Point of the Mountain congestion, as well as the stretch from north Orem to American Fork, and the I-15/215 interchange. As the graphic shows, there is usually quick traffic on I-215 or I-80.

DESIGN Jonathan Hull   COUNTRY/REGION   The USA

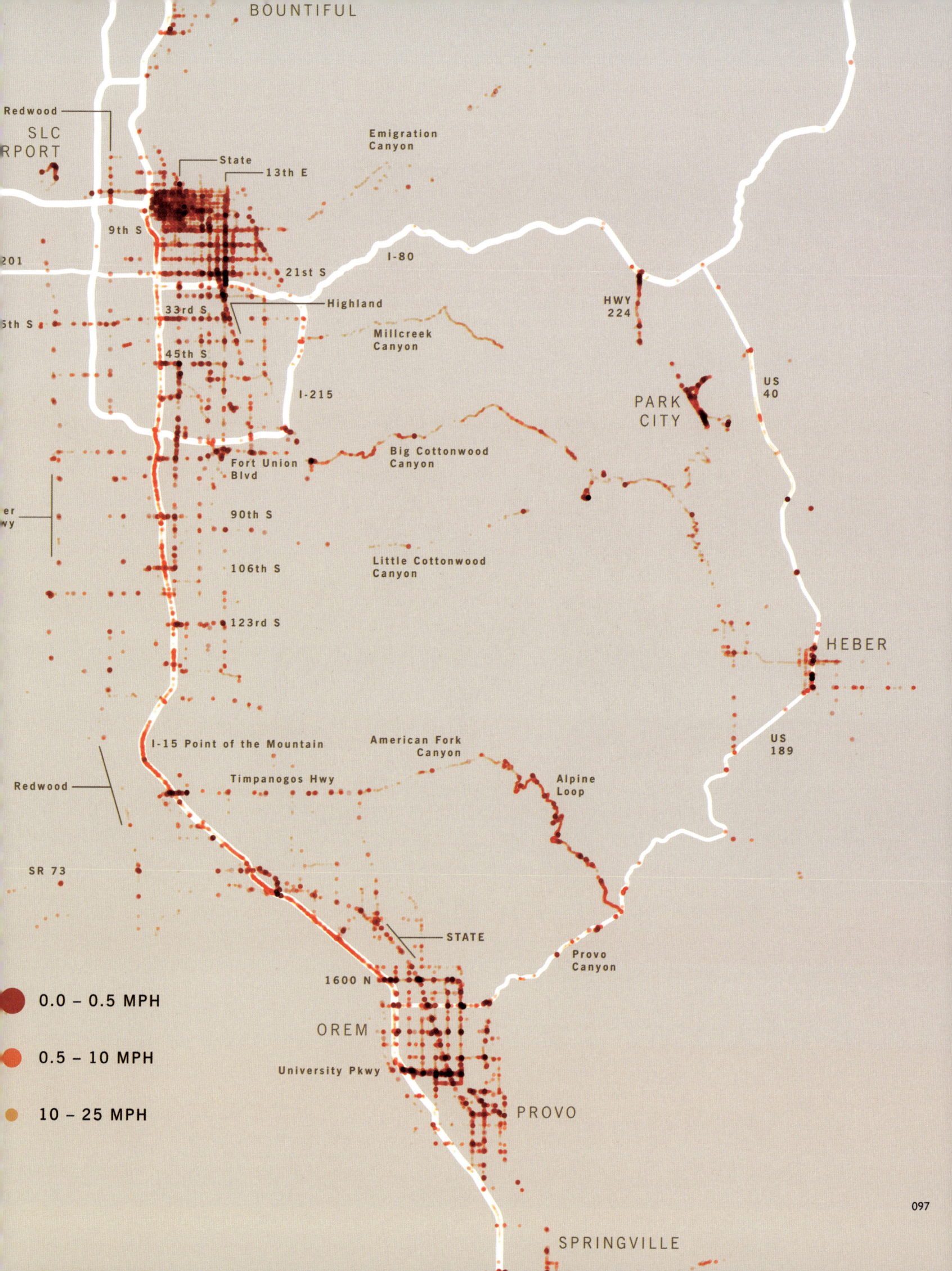

MAPPING THE PHYSICAL ENVIRONMENT

Moscow Metro Map

Named Moscow Ant Hill, this art piece is an exclusive Moscow metro map made of natural materials. Representing the lines of the metro system by real birch branches the designer expresses a metaphor and his nod to all the "ants"—the megalopolis citizens, who have built this beautiful city, metro and life!

DESIGN Raushan Sultanov   COUNTRY/REGION Russia

MAPPING THE PHYSICAL ENVIRONMENT

London: the Capital of Romance

This mosaic illustration was for the 2012 Saint Valentine's special issue of Evening Standard magazine. To correspond to this theme, a red heart was created on the center of this map collaged by fractional photos taken in London. The grid on which this mosaic is based is actually a very accurate map of the London metropolitan area.

DESIGN Charis Tsevis   COUNTRY/REGION Greece

Botanical South America

Botanical South America is an info-graphic illustrated map made to characterize and capture the immense landscape, jungle, and botanical life that thrives on the continent of South America. All the different plant life specific to each South American country is neatly organized on the map. To emphasize the plants and add to the character of the image, Civiello incorporated random edges to the lines of the outside of the continent, which gave the map an organic and natural shape as plants might have, while also maintaining the actual shape of the continent.

DESIGN Dom Civiello   COUNTRY/REGION The USA

The City of London Illustrated Map

The design of this map of the historic centre of London is inspired by decorative antique map prints and the map art of MacDonald Gill (1884-1947). The coat of arms at the top centre is the official arms of the City Of London, a historic administrative entity separate from the rest of the metropolis, while along the left and right edges are twelve shields representing the 'Great Twelve' Livery Companies, or historic trade guilds. The border decoration is inspired by classical architecture, featuring two Corinthian columns supporting a lintel; this makes reference to both the original Roman founders of the city, and the neo-classical architecture that abounds: the Bank of England and St. Paul's Cathedral being two prominent examples.

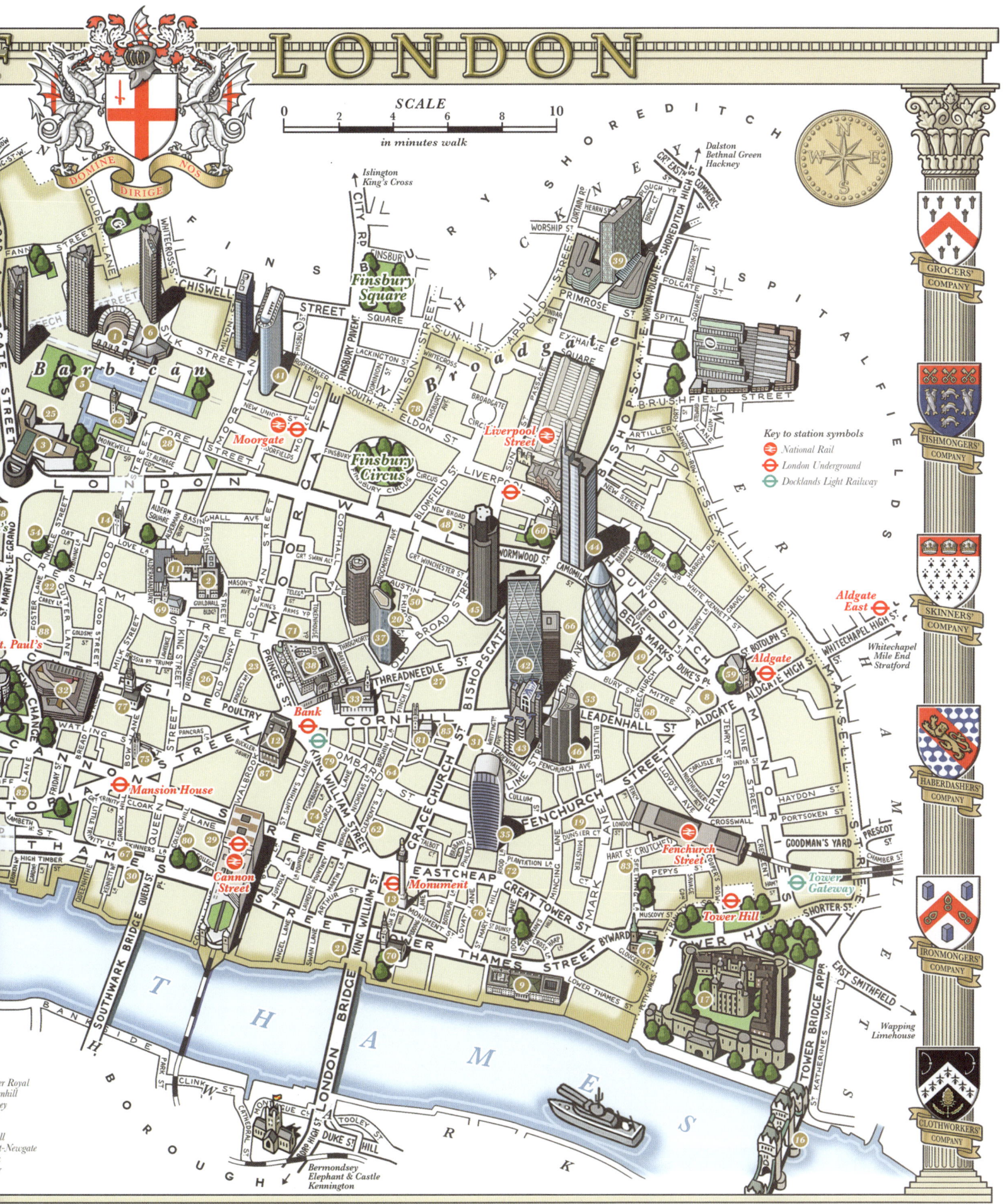

An Illustrated Map of Laurie Lee's Journey Through Spain

This map illustrates the journey taken by the English writer Laurie Lee (1914-1997) across Spain between 1935 and 1936, recounted in his memoir "As I Walked Out One Midsummer Morning" (published 1969). This tale of a young man's solo journey on foot across Spain in the months before its devastating Civil War is a fascinating read, vividly describing a poor agrarian society that seems unrecognizable today.

The design includes illustrations of places that Lee visited along his journey, including Vigo, Valladolid, Toledo and Seville. These were drawn to appear as they would have in the mid 1930s. The border was inspired by traditional Spanish tile decorations.

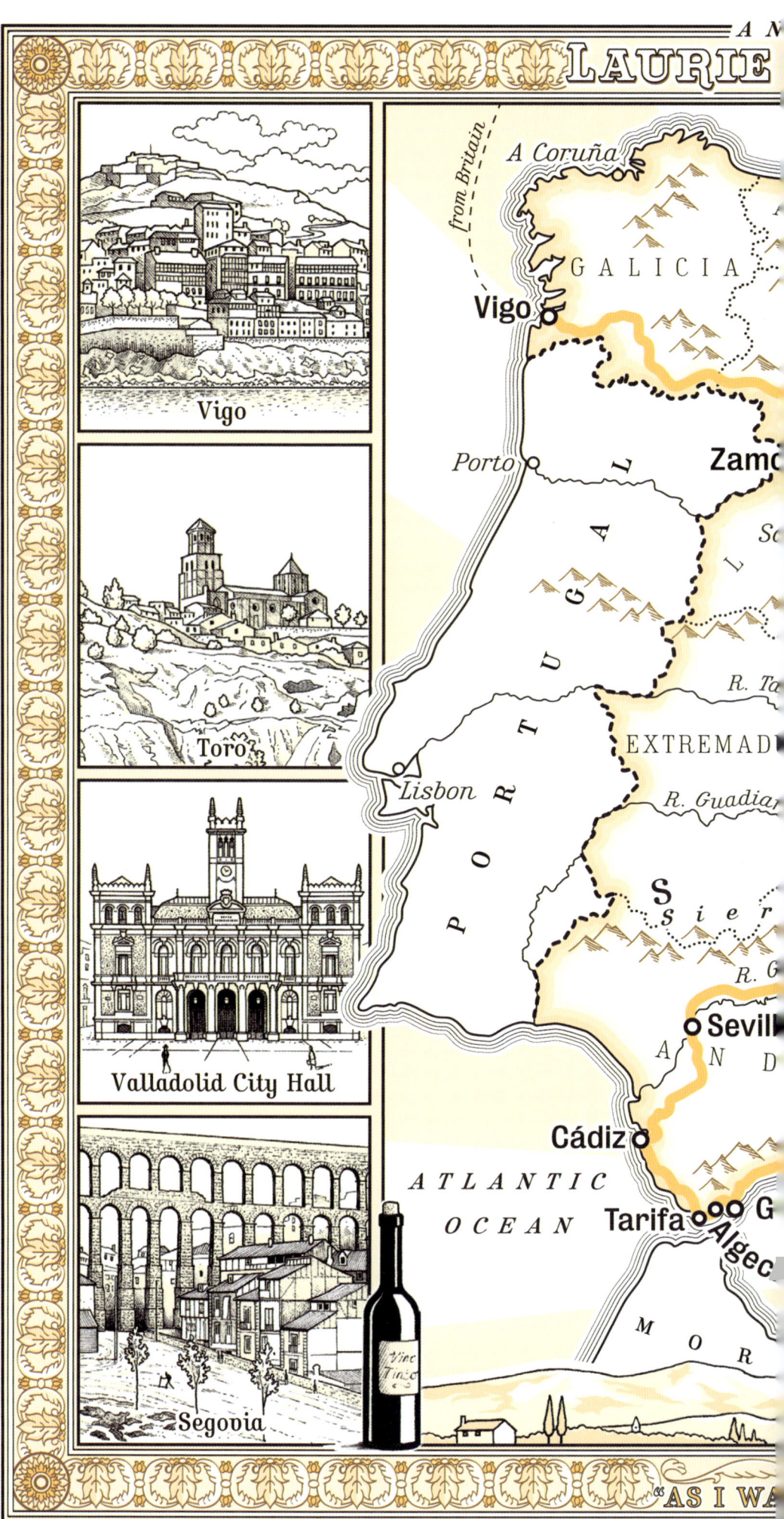

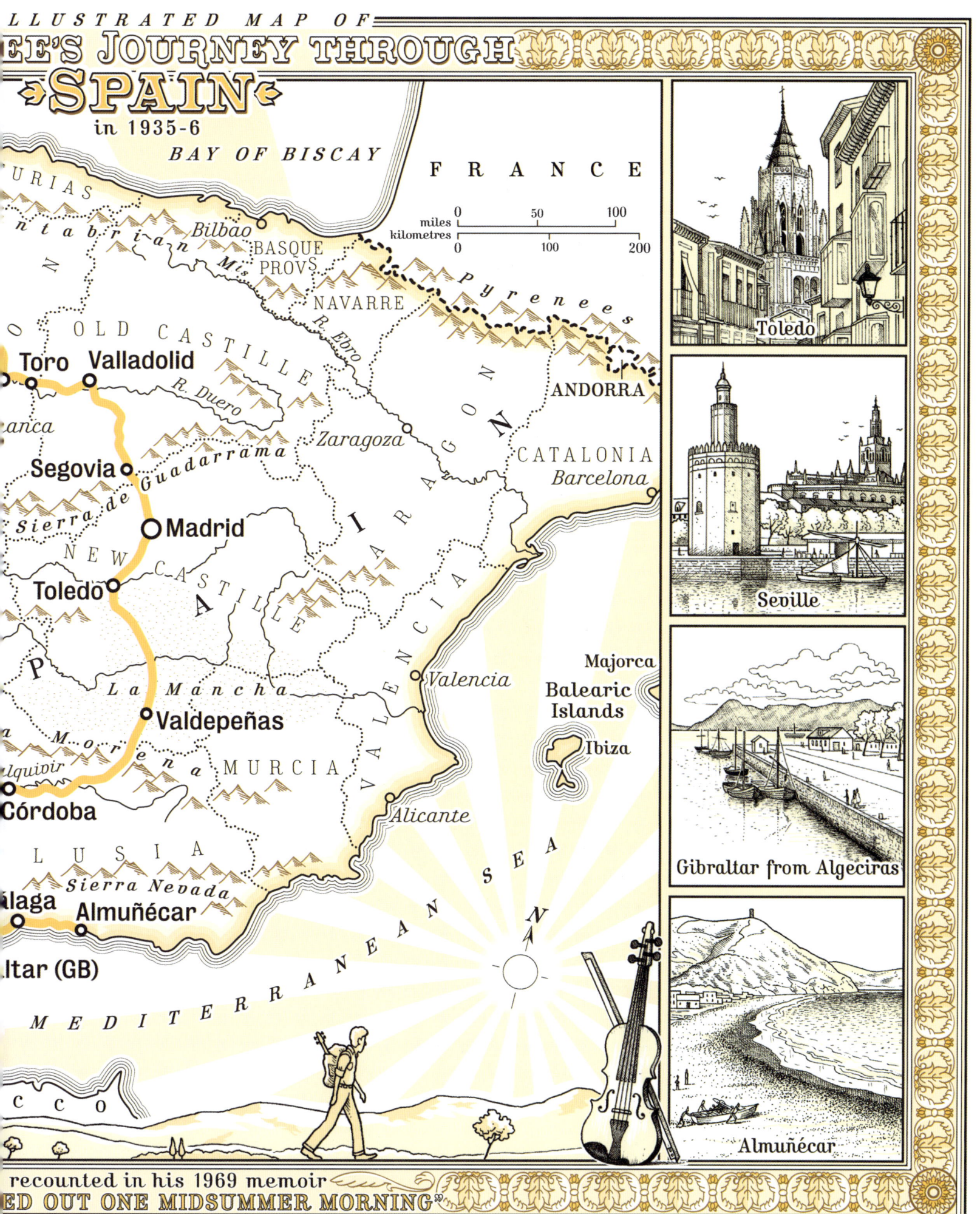

Illustrated Map of Central London

This design was inspired by decorative antique map prints and the map art of MacDonald Gill (1884-1947). The map is embellished with illustrations of various famous and prominent landmarks in elevation, including major museums, churches, skyscrapers and attractions. Locations of all railway stations, major termini and London Underground stations are labelled.

The border decoration features coats of arms of the eight boroughs that appear on the map (Camden, City of Westminster, Hackney, Islington, Kensington & Chelsea, Lambeth, Southwark and Tower Hamlets) and the City of London (a separate administrative entity). Four illustrations in each corner show scenes and locations in different parts of the capital.

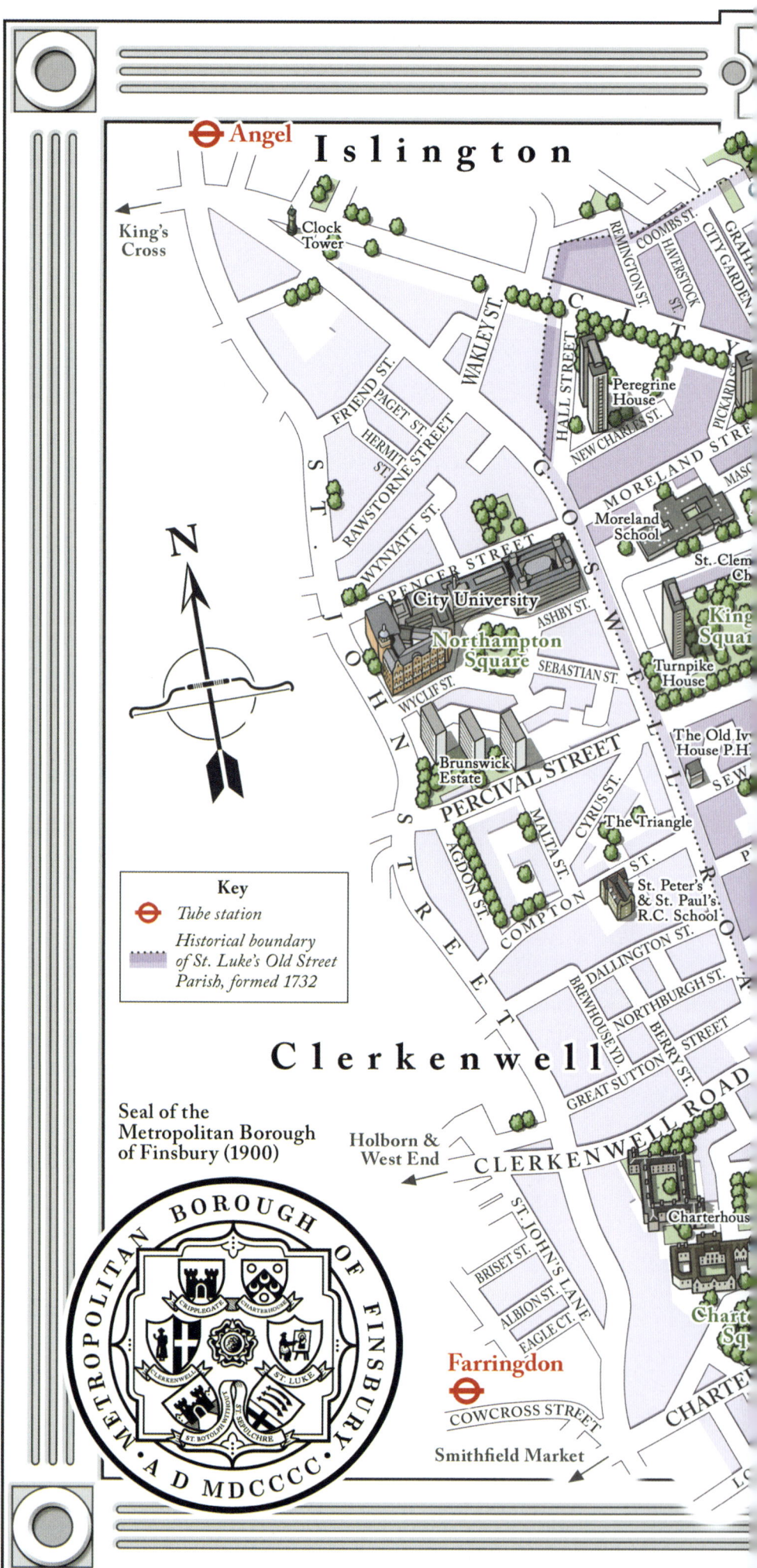

The Neighborhood of St. Luke's, London EC1

St. Luke's Trust is a charitable organization based in the St. Luke's neighborhood in east-central London. Hall was commissioned by this organization to design a decorative map of their neighborhood for use in their publicity and to sell as a print.

The intention was to raise the profile of this mostly residential neighborhood, located between the City, Angel, Clerkenwell and Shoreditch, which is little known to most people nowadays and is commonly referred to simply as 'Old Street' owing to that station's location nearby. The design includes the coat of arms of the local borough of Islington, and the seal of Finsbury, the old borough that was incorporated into Islington in 1965.

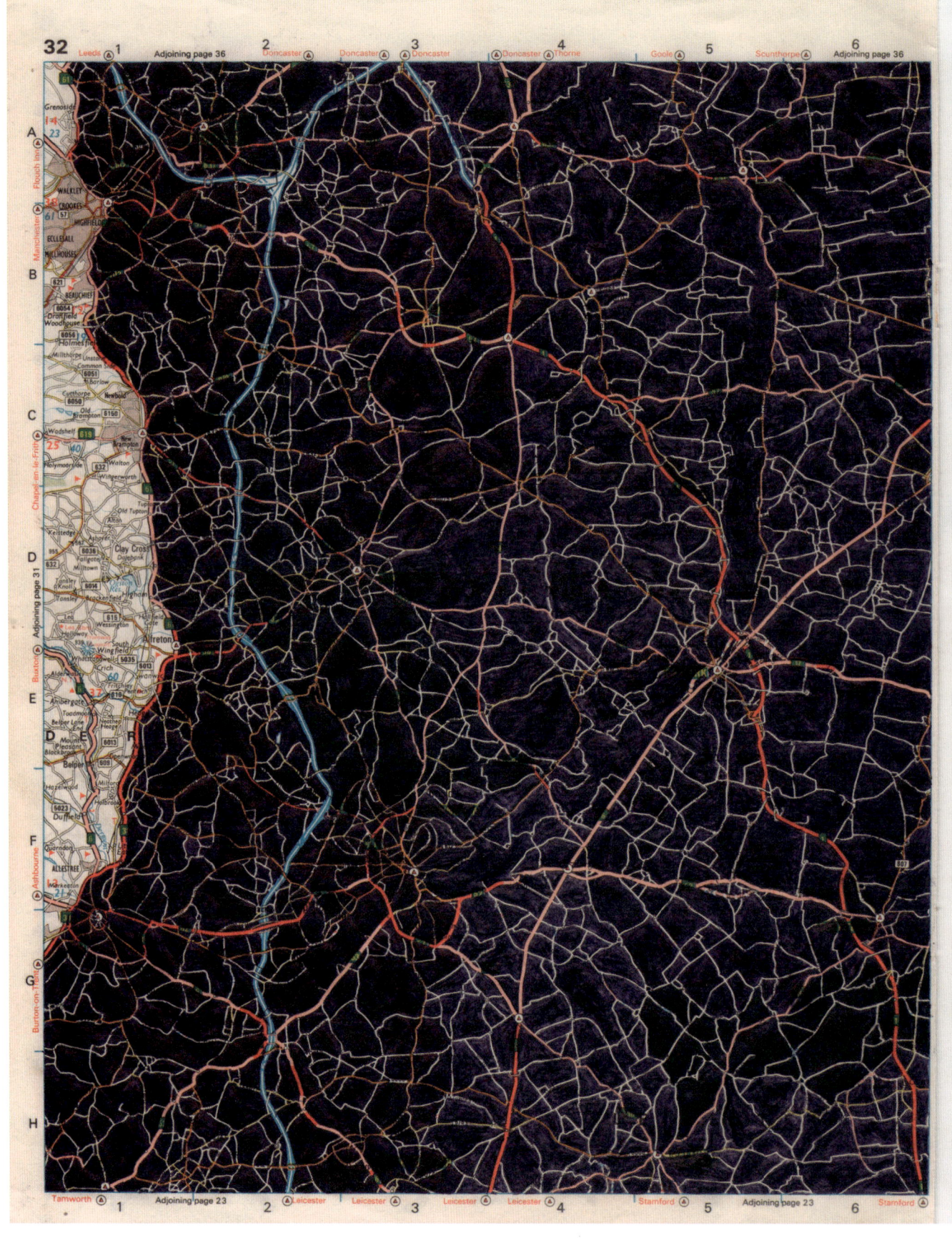

Blue Roads Map

A lot of the time nowadays we are overloaded and overwhelmed by so much information and things wanting our time and energy. This concern inspired Norman to use an everyday blue biro pen to draw in the land areas of a map between the roads. In this way, she tries to convey the beauty and importance of not knowing, of being quiet and finding your own path.

DESIGN Louise Norman   COUNTRY/REGION   The UK

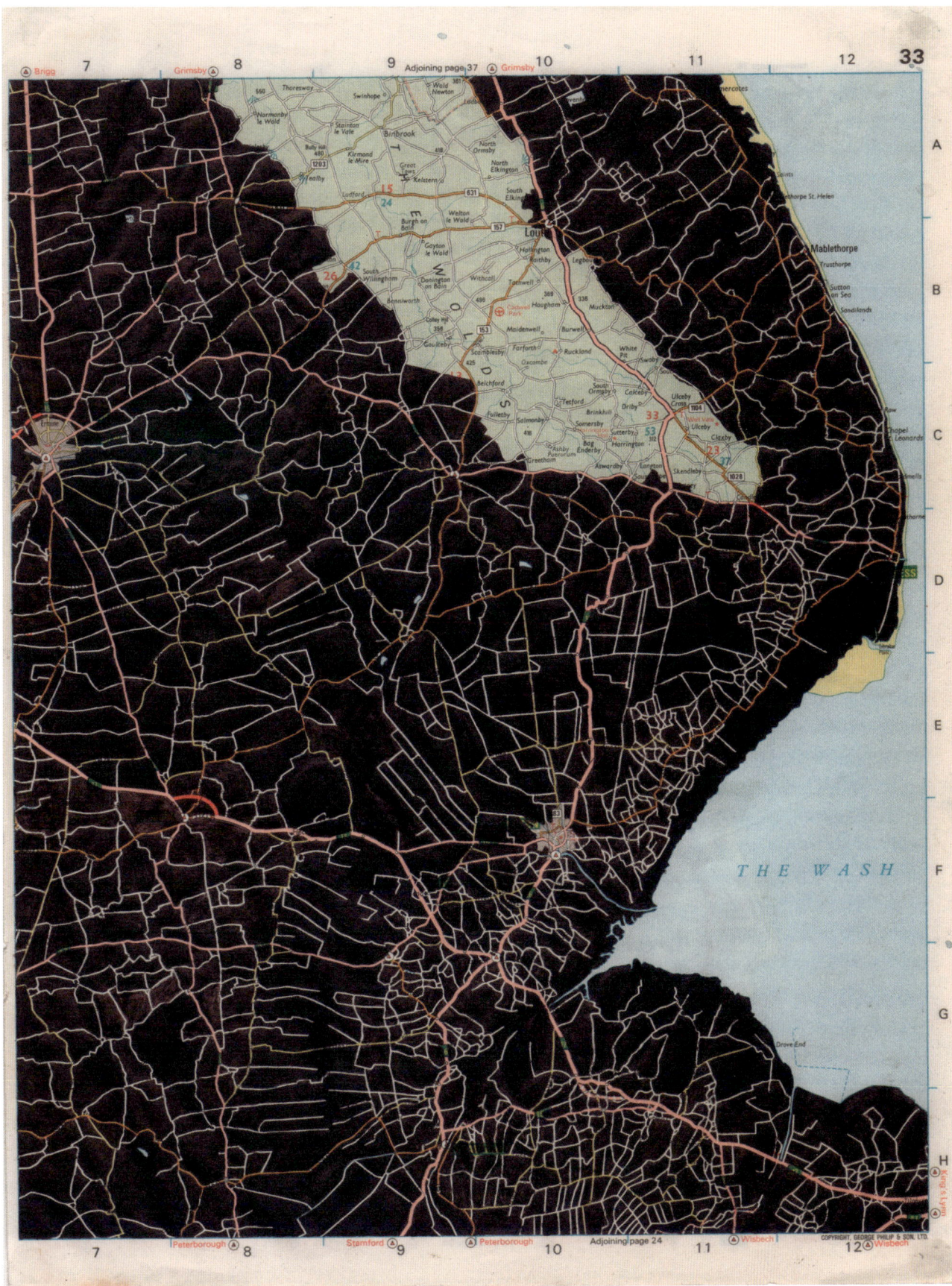

The Wash Roads Map

Norman used an everyday black biro pen to draw in the areas of land between the roads of a map and just leaving the green spaces free. Using a repetitive process of drawing in each area blocks out all information and creates a new simpler map reflecting her hand/mind and the process used to achieve it. Also Norman have left the green spaces as they are to evoke the wildness of nature and how nature will always try to reclaim a space or map.

DESIGN Louise Norman   COUNTRY/REGION   The UK

Time & Tide 2015

Composed of multi-layered hand-cut worlds, this work shows different elements such as currents and tides, tracks showing ferry routes and other vessels. Whilst the land-masses overlap and shift from layer to layer, the different tracks interweave and suggest new complex tracks and patterns.

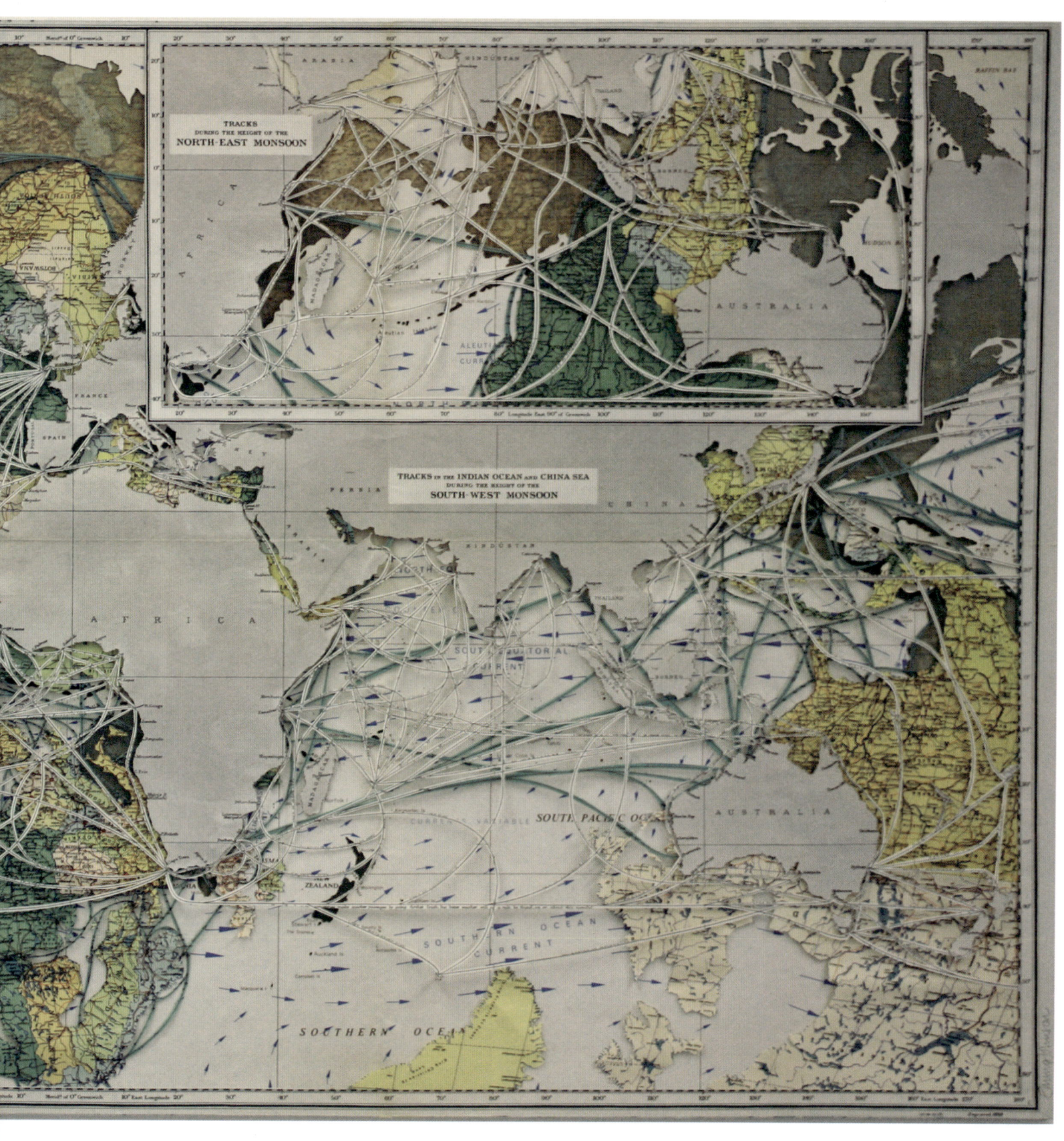

DESIGN Emma Johnson COUNTRY/REGION The UK

Manhattan Transit 2014

This work interprets a multi-layered hand-cut map of Manhattan. Cutting through the formalised grid-type structure of the city by superimposing, inverting and subverting its structure, a new complex grid structure emerges—still recognizable, but gridlocked.

DESIGN Emma Johnson   COUNTRY/REGION   The UK

Dislocation: Time & Place 2011

The paper-cut map is composed from places the artist has lived throughout her life, mapping her own personal journey. Its dissection and "dislocation" represent the fragmented nature of memory, time and history as well as a celebration of the intricate aesthetics of mapmaking.

DESIGN Emma Johnson   COUNTRY/REGION   The UK

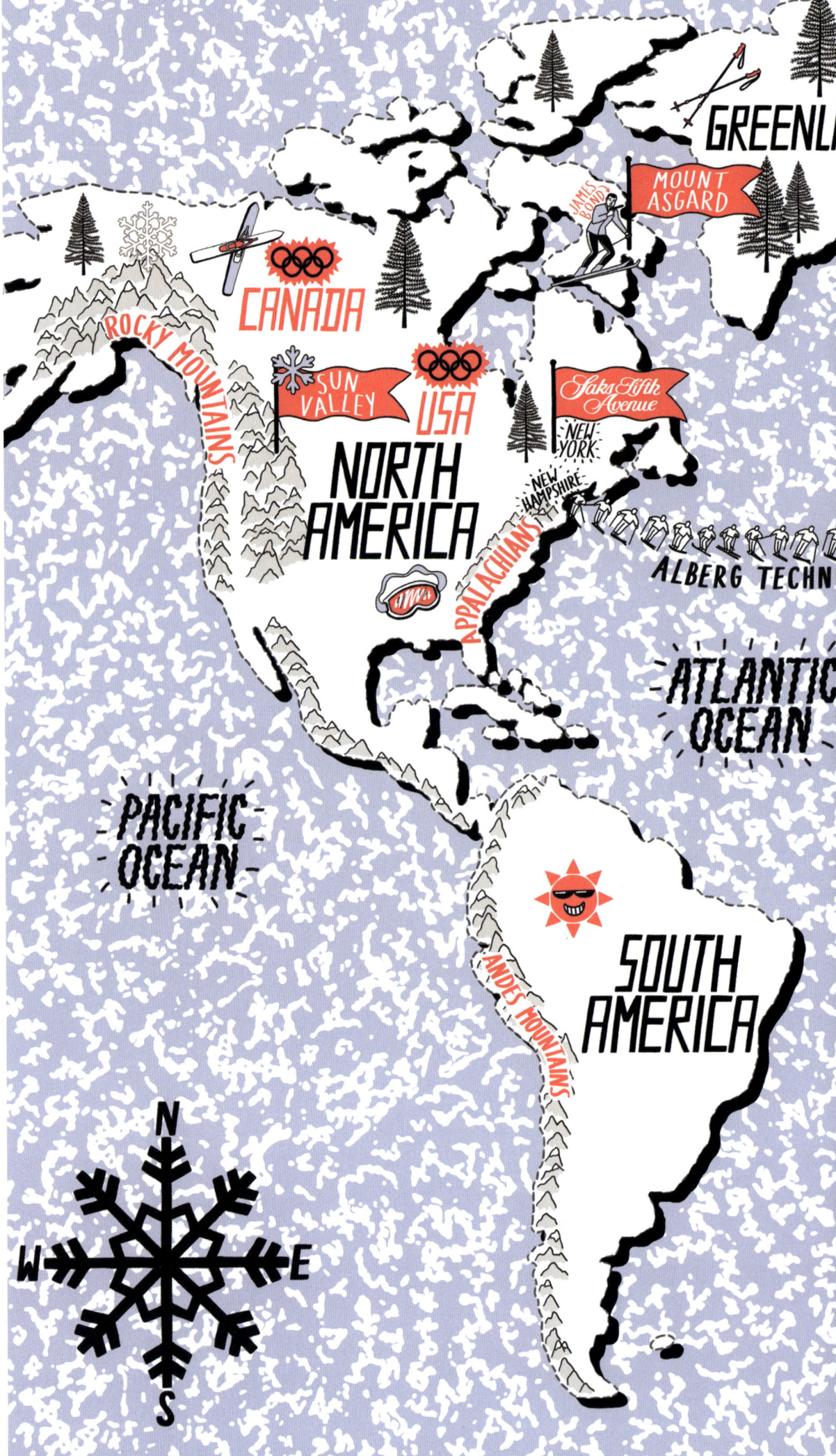

Moving Mountains Map

This is a world map about skiing and ski fashion, which was a part of the 'Moving Mountains' Winter exhibition at Somerset House. The small symbols on the map represent different events and ski fashion items that are associated with the sport, as well as locations of several Winter Olympic Games. The design of the map was meant to be youthful and fresh.

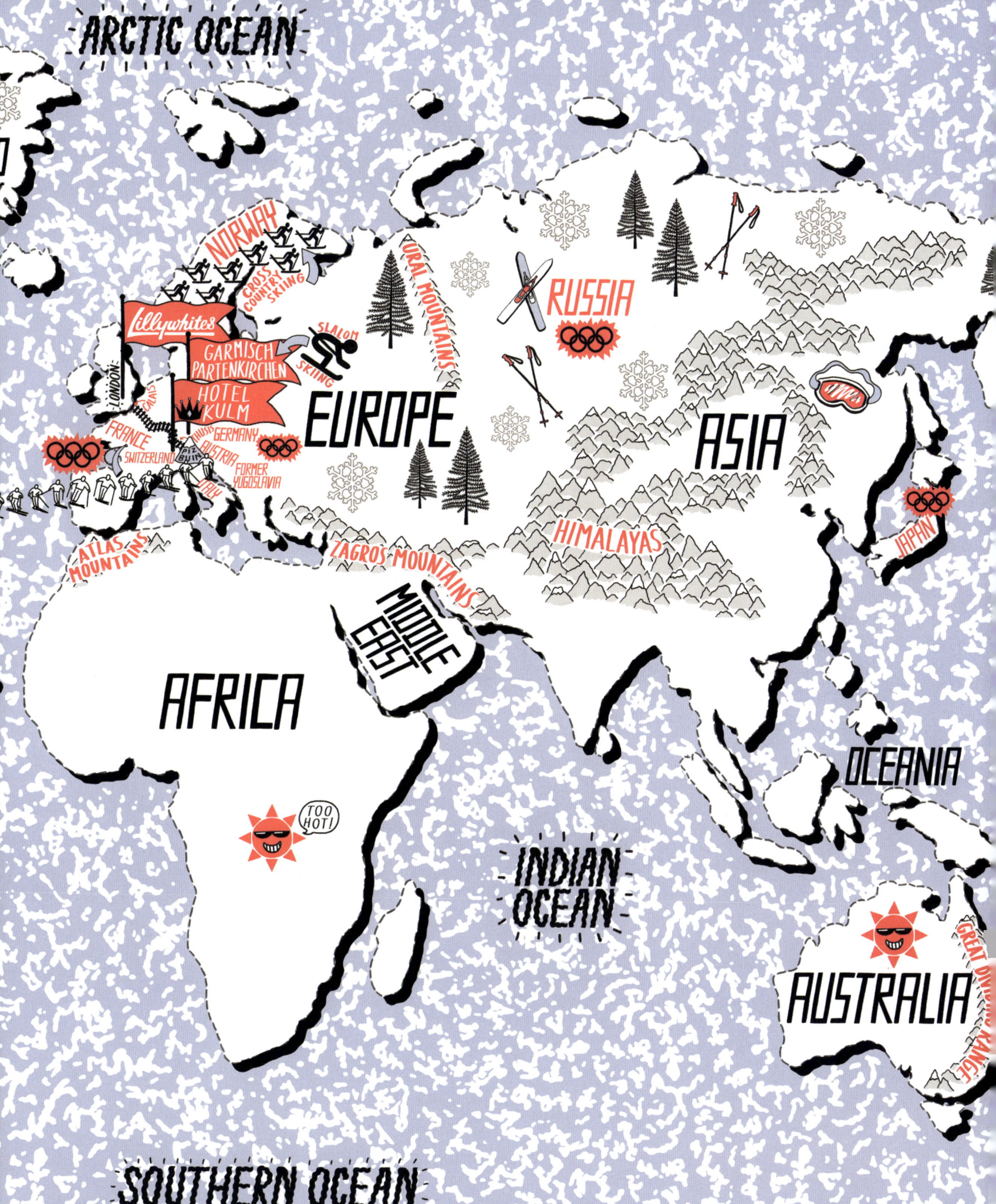

MAPPING THE PHYSICAL ENVIRONMENT

The Greatest Adventure

The brief of this project is to provide a solution for people who want to visit and explore Bandung, West Java, Indonesia by introducing Bandung's tourist attraction especially its thematic parks through an illustration map. The thematic parks in the city take huge part in designing this map. The map shows the routes connecting the three thematic parks. The iconic and nearby places are essential for the map as well. The names of the locations are shown in the legend. The characters and icons of this map are inspired by Wes Anderson's piece, Moonrise Kingdom.

DESIGN Owi Liunic   COUNTRY/REGION Indonesia

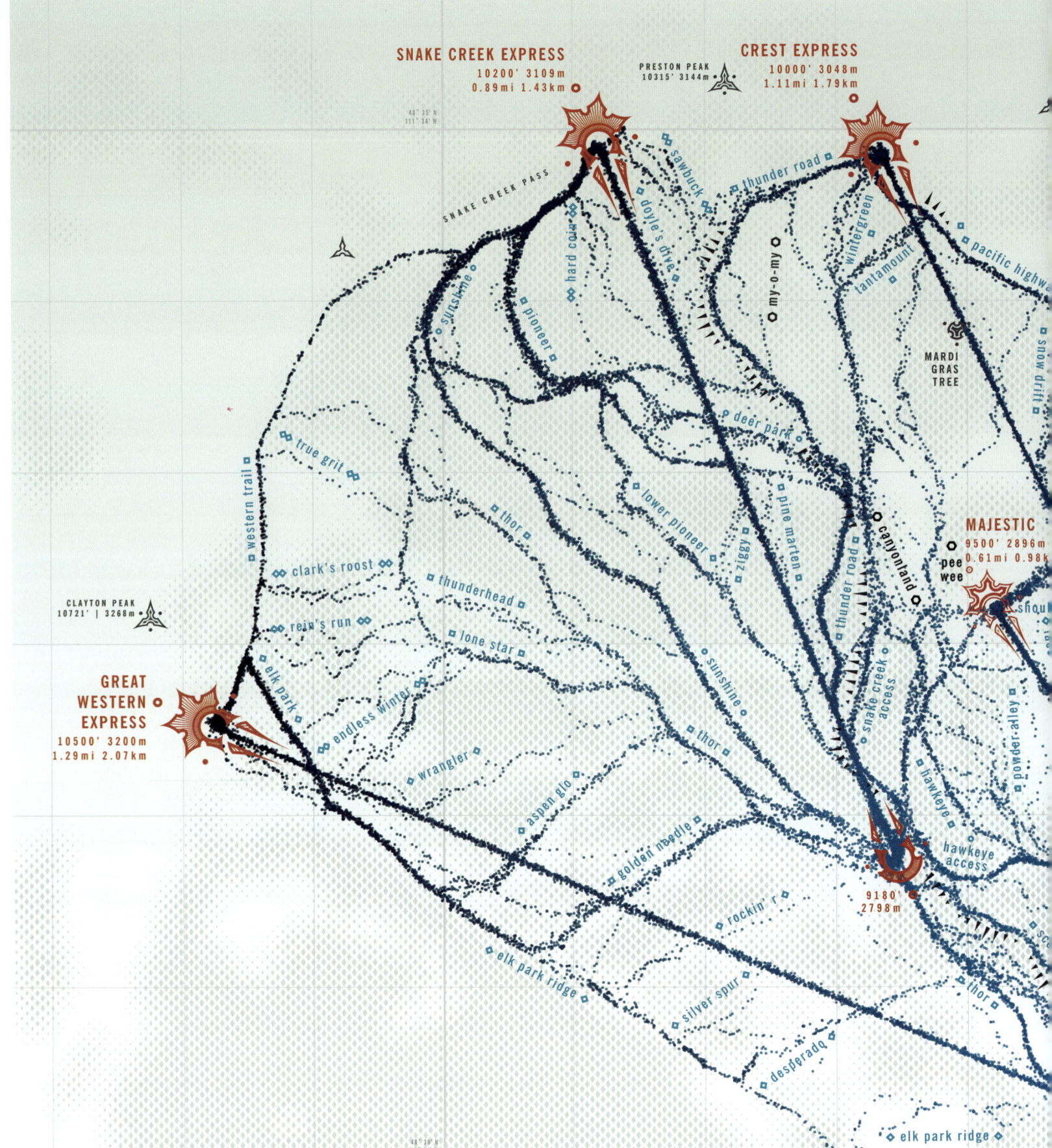

Brighton Resort Trail Map

This map was created via GPS, mapping the resort trails at Brighton Resort, Utah over multiple seasons. As far as a mapmaking project, the first task was reconciling the resort's trail map with the reality of the mountain and trail markers. The designer also created complex markers for the lifts as well as a patterned background based on the forest at the resort.

Dublin City Map

This is a large tourist wall map of Dublin City for the Dublin Visitor Centre. The map aims to be an informative visitor's guide as well as an attractive piece of art which would encourage people into the store. A palette featuring Georgian colours was used to reflect some of the city's architecture. The designer also introduces the appeal of the city into the map by illustrating the buildings and cultural attractions in a light-hearted way.

BAILE ÁTHA CLIATH BY PETER DONNELLY

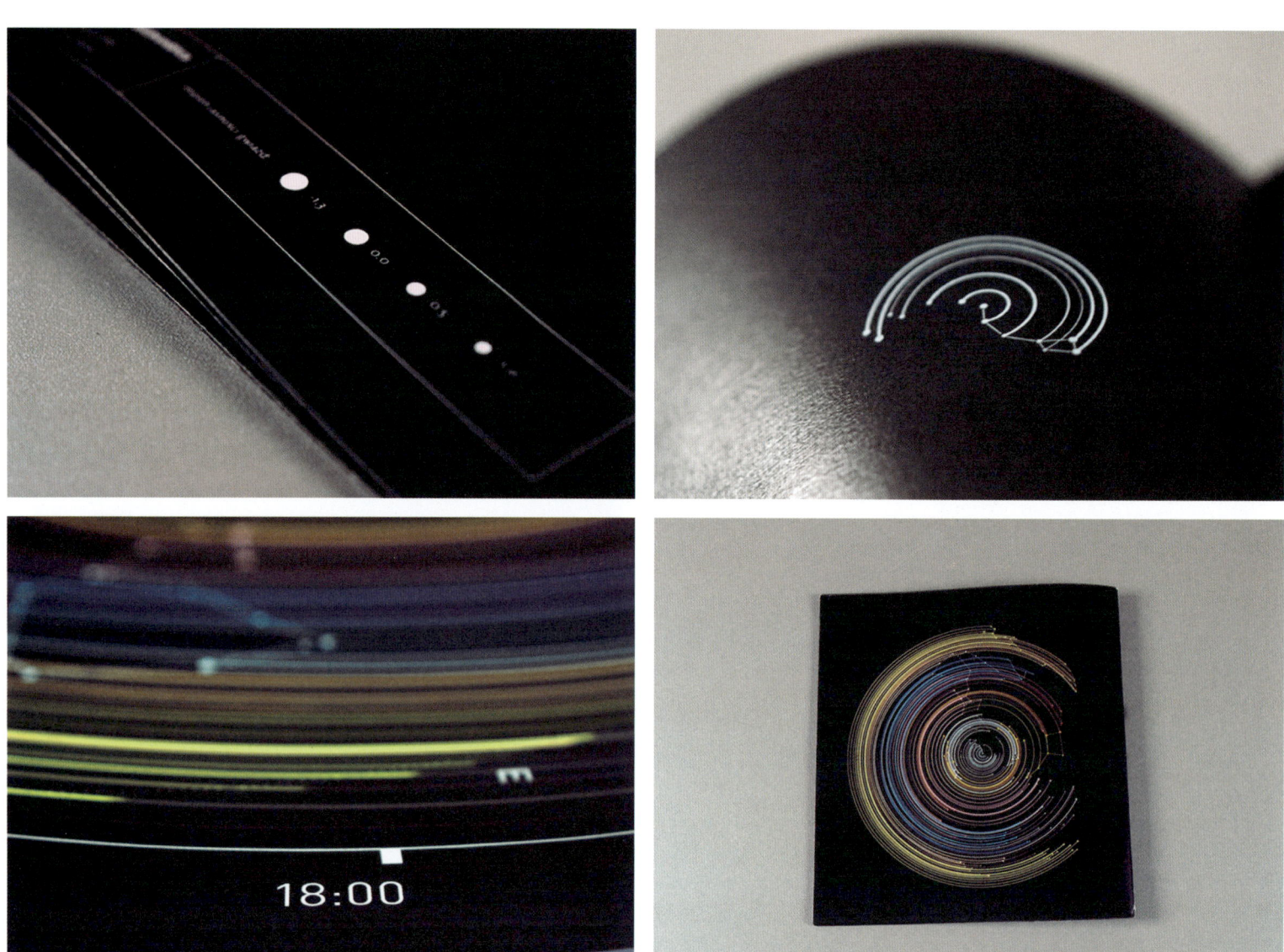

MAPPING THE PHYSICAL ENVIRONMENT

Sky Map

Sky map is an educational infographic visualizing apparent motion of the stars, during the longest night in the year (the Winter Solstice in the northern hemisphere, lasting almost 17 hours from around 3 pm on December 23rd to around 7 am on December 24th). This particular night sky shows the longest journey of visible stars above our heads. The project was based on source data provided by the Planetarium in Silesia, a scientist and astronomical centre in Poland. The sky was observed from Silesia Province, Poland.

DESIGN Paulina Urbańska   COUNTRY/REGION   Poland

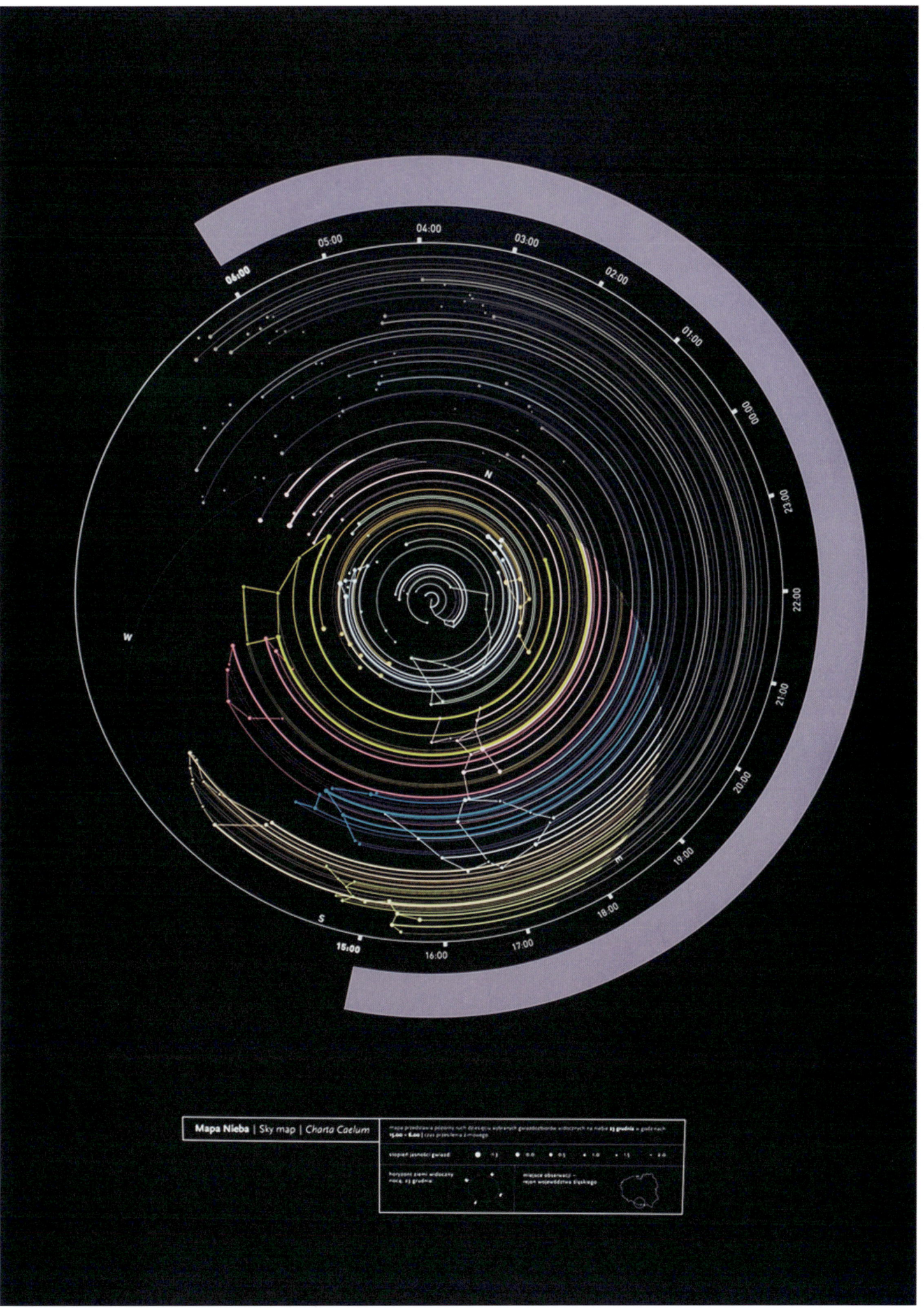

Guide to Saraburi

The illustration was meant to be a guide for tourists in Saraburi, showing the landmarks of the city. Yeukprasert created the icons by simplifying the renowned landmarks in a graphic form for easy recognition.

DESIGN Chinapat Yeukprasert COUNTRY/REGION Thailand

Shauna

Shauna is an inlaid map collage of Colleen Applegate (1963–1984), the deceased pornographic star known as Shauna Grant. Certain areas of Los Angeles, such as the San Fernando Valley, have become geographic epicenters for the production of pornography. The map collage of Shauna reflects upon the relationship between persona, pornography, and place and how it has redefined specific cultural regions.

DESIGN Matthew Cusick COUNTRY/REGION The USA

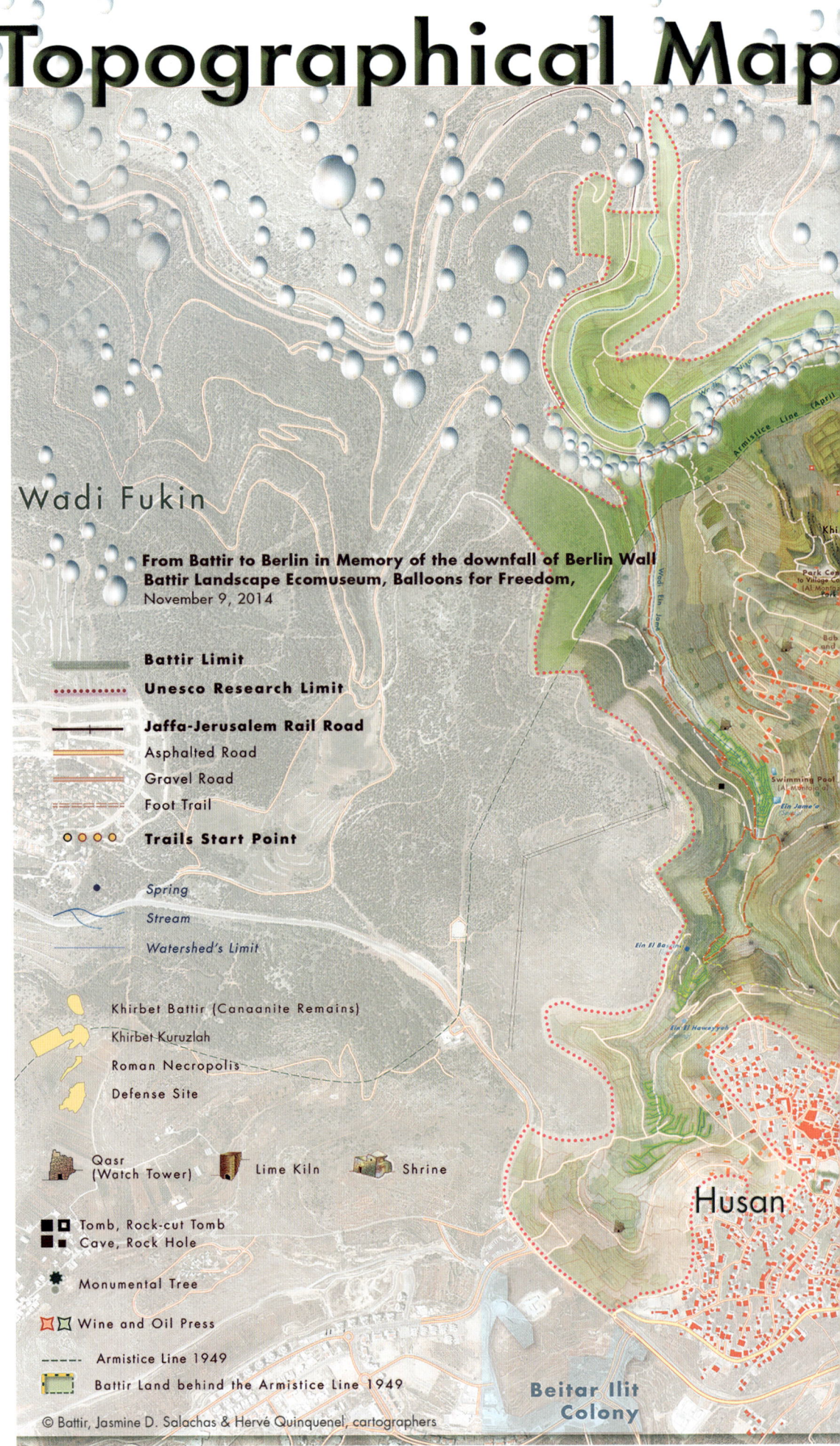

Battir Map

A UNESCO World Heritage site, Battir is a Palestinian village located in the West Bank, on the 1949 Armistice Line. Since 2007, a project has been carried out by a multidisciplinary team of Palestinian professionals to explore the characteristics of this 12 km² territory. These topographical maps are the achievements of the project. They are useful in studying the site and bringing international attention to its recognition and protection.

Connect People To People

**UNESCO RAMALLAH OFFICE in partnership
with BATTIR VILLAGE COUNCIL**

In cooperation with:
THE MINISTRY OF LOCAL GOVERNMENT
THE MINISTRY OF TOURISM AND ANTIQUITIES
THE GOVERNORATE OF BETHLEHEM
BATTIR BALADNA (LOCAL COMMUNITY COMMITTEE)

**BATTIR LANDSCAPE
CONSERVATION AND MANAGEMENT PLAN**

PROJECT COORDINATION AND WORKING TEAM SUPERVISION:
  Arch. Giovanni Fontana Antonelli

LANDSCAPE SURVEY, GIS MAPPING AND ADVOCACY PLANNING:
  Arch. Samir Harb
  Arch. Mohammad Hammash
  Eng. Hassan Muamer (from February 2010)
  Arch. Mohammad Abu Hammad (until September 2009)

LANDSCAPE PLANNING SCIENTIFIC ADVISOR:
  Arch. Pasquale Barone

CONSULTANCY FOR GEOMORPHOLOGICAL ASPECTS:
  Mr. Francesco Cini

COMMUNITY PARTICIPATION:
  Dr. Claudia Cancellotti
  Dr. Patrizia Cirino
  Dr. Nicola Perugini

Legend

- Battir Limit
- Unesco Research Limit
- Jaffa-Jerusalem Rail Road
- Asphalted Road
- Gravel Road
- Foot Trail
- ○○○○ Trails Start Point
- TRAIL 1
- TRAIL 2
- TRAIL B
- TRAIL C
- • Spring
- Stream
- Watershed's Limit

- Khirbet Battir (Canaanite Remains)
- Khirbet Kuruzlah (Roman Castrum)
- Roman Necropolis
- Defense Site

- Qasr (Watch Tower)
- Lime Kiln
- Shrine
- Tomb, Rock-cut Tomb
- Wine and Oil Press
- Cave, Rock Hole
- Monumental Tree
- Monumental Stone, Rocks
- Terraces, Retaining Walls
- Cliff
- Highvoltage and Electric Line

- Armistice Line 1949
- Battir Land behind the Armistice Line 1949

0   300   600 m
Scale : 1:15 000

Summer 2013 in Battir: Ali Muammar, Itidal Muammar, Ehab Muammar, Bader Al Buthma -Battir. And Sylvain Gonnet -Surveyor, France
April/May 2014: New Survey by Hervé Quinquenel -Cartographer/GIS-Eng. France- Orienteering Map Collaborative Project of Battir
Sept.17, 2015: Mapping and Coordination since May 2012, Jasmine D. Salachas - Cartographer, les Cafés-cartographiques

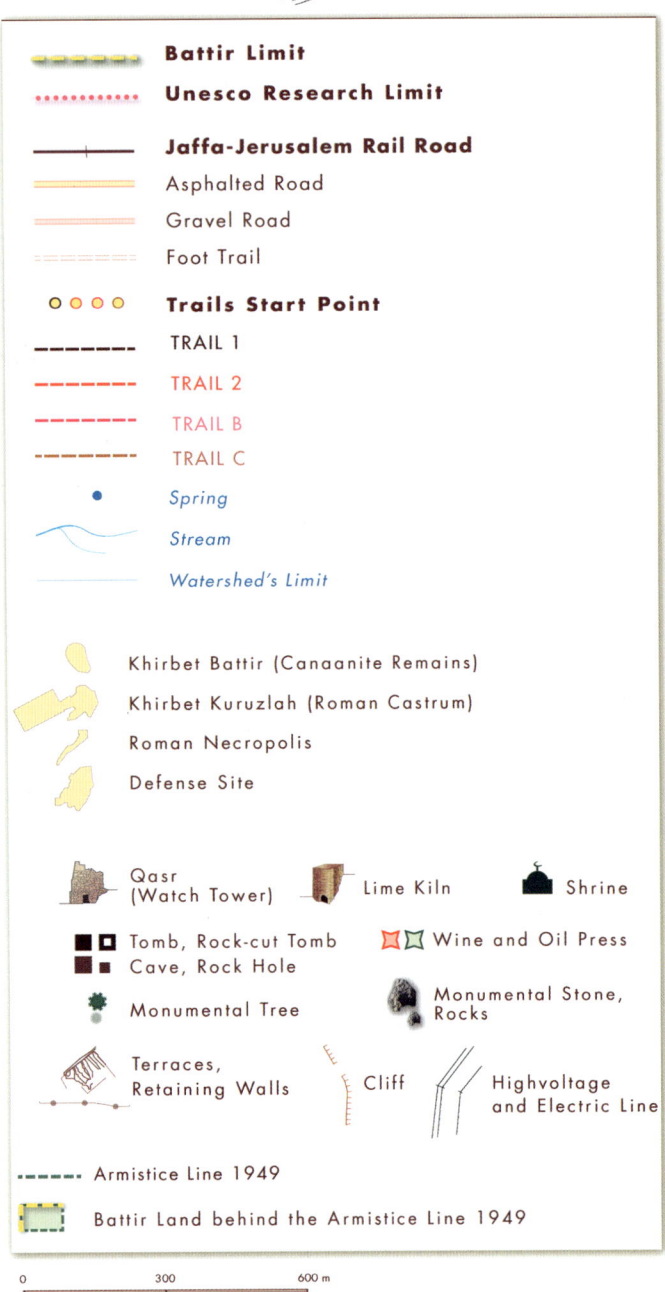

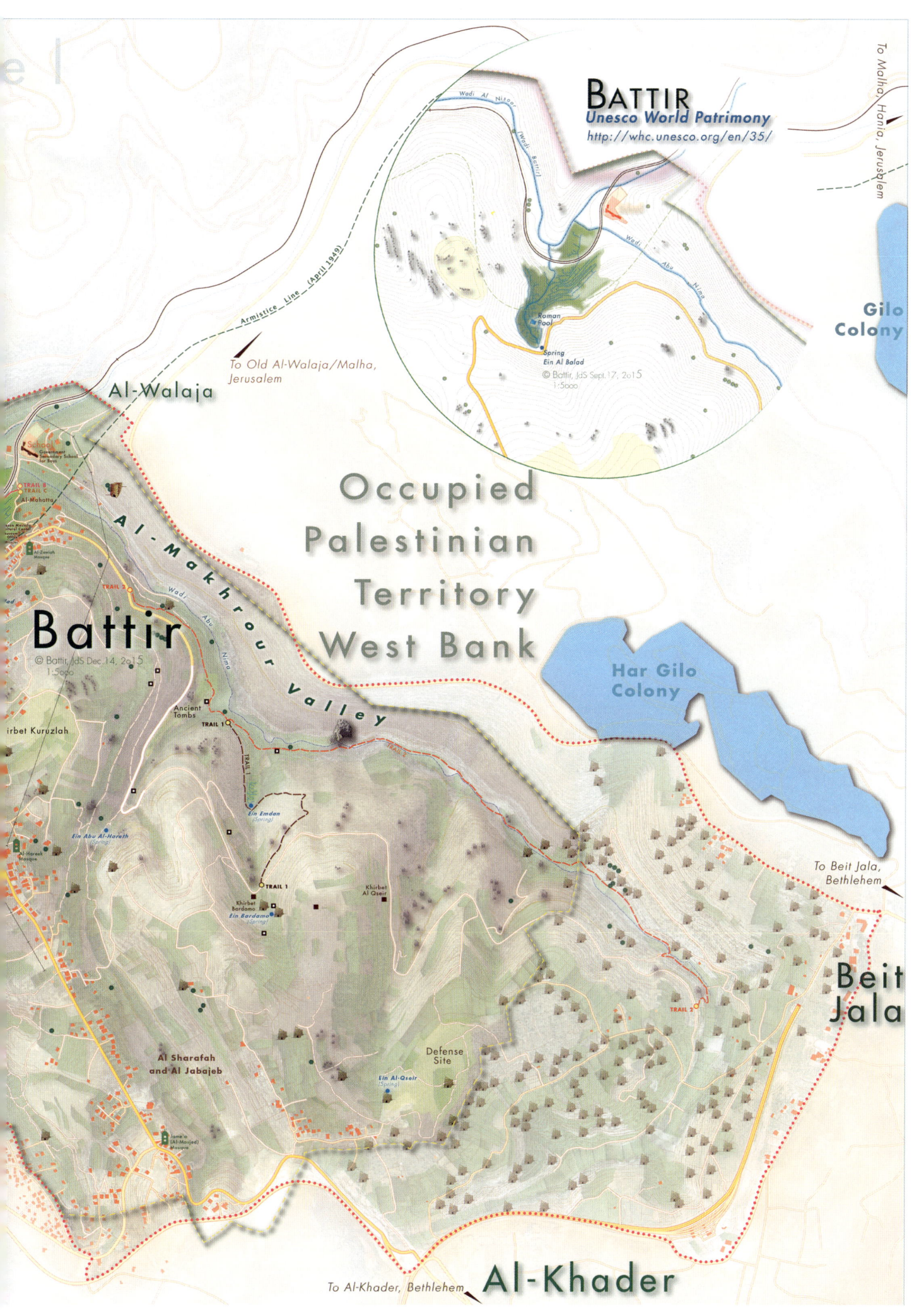

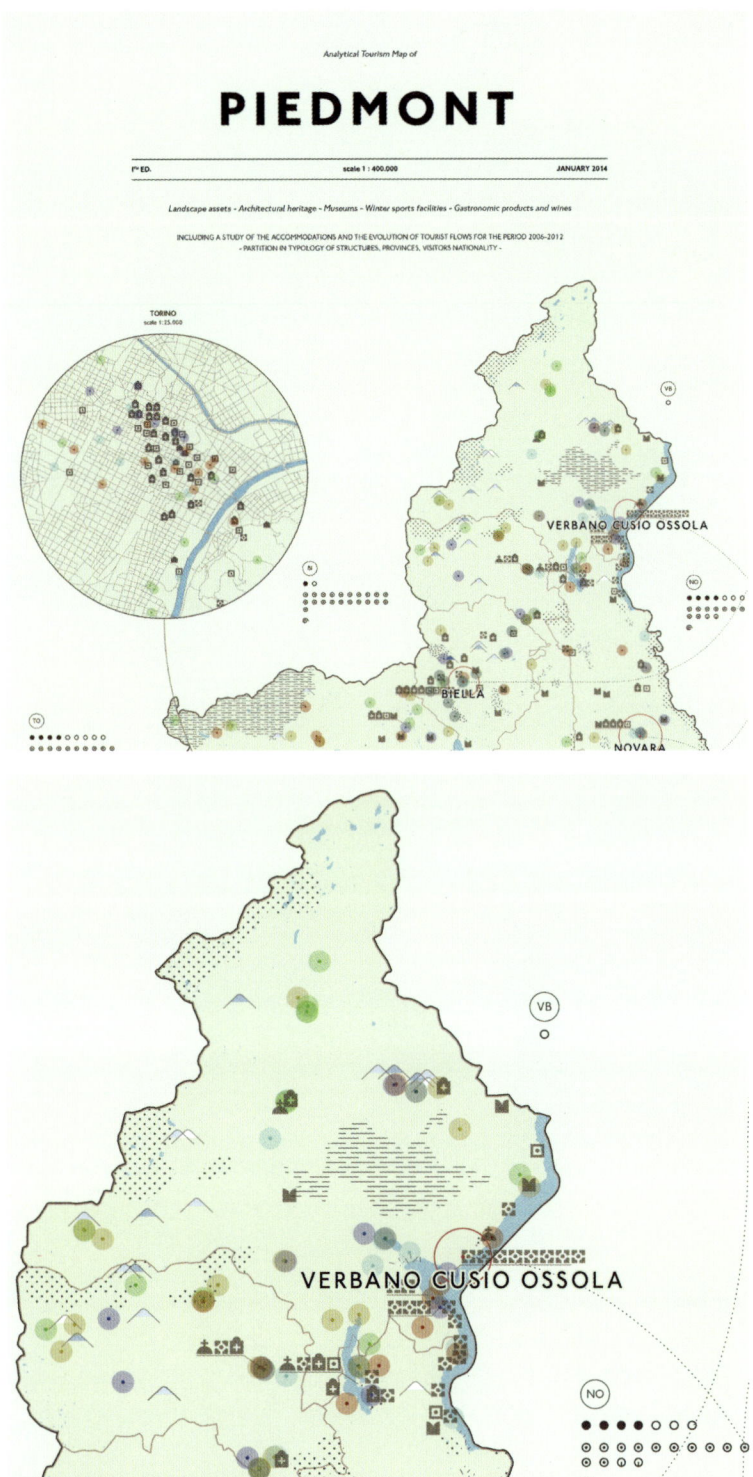

Data extraction by ItaliaNLP Lab.

The Analytical Tourism Map of Piedmont

Based on open data, the Analytical Tourism Map of Piedmont aims to investigate the tourism resources in a complete and detailed way, considering both the attractions of the region and the evolution and dimension of tourist flows and accommodations. On the left side of the poster, the Piedmont map shows the most important architectural heritage buildings, museums, winter sports facilities, gastronomic products and wine, etc.. Every element is geolocated on the map, and it's easy to identify the beauties of different areas and the best places to visit. On the right side, the visualization shows the categories of tourist flows and compares the accommodations distributed throughout the 8 provinces of Piedmont between 2006 and 2012.

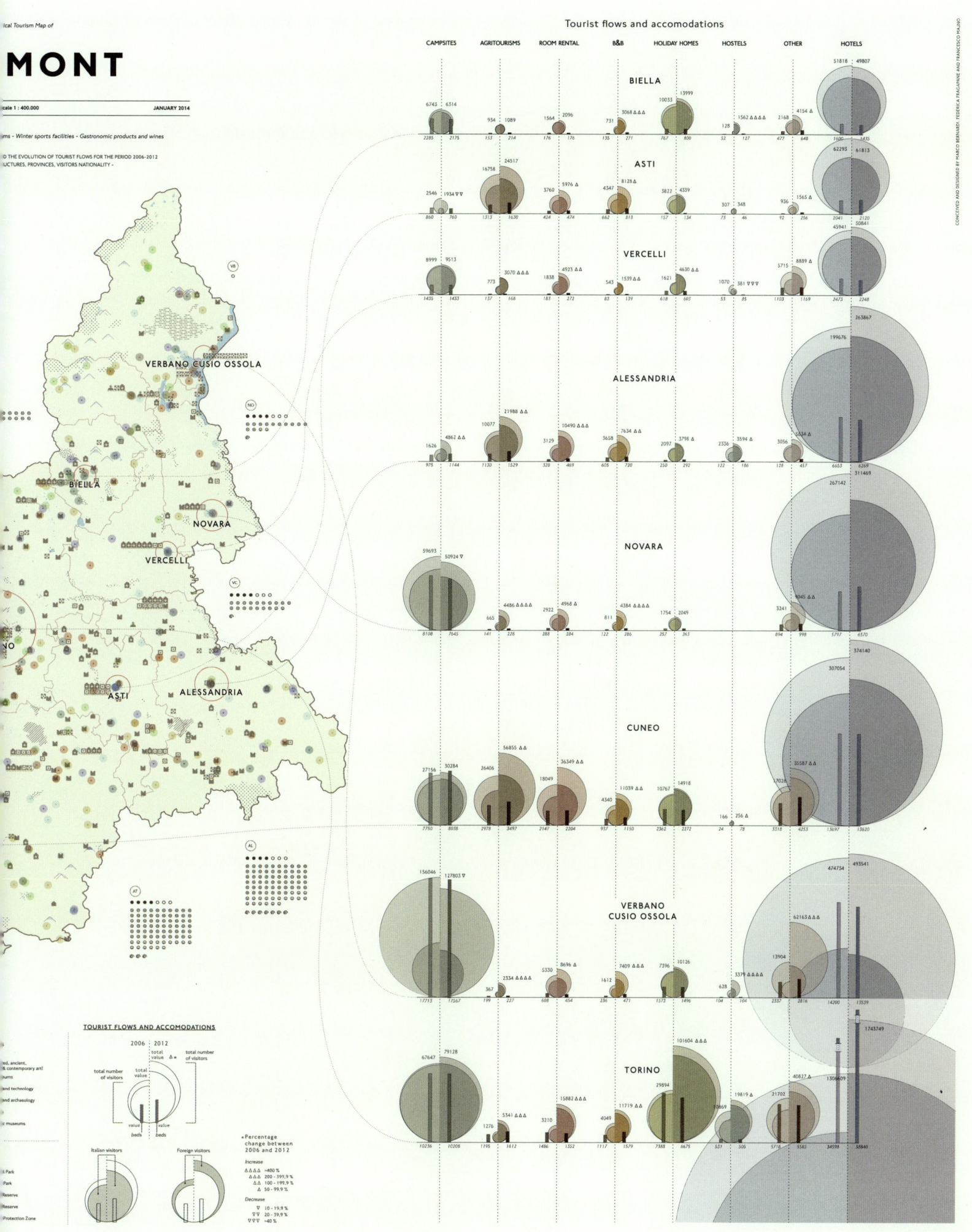

MAPPING THE PHYSICAL ENVIRONMENT

The Mixmaster Series

In these three Map Works by Matthew Cusick, the artist captures the beauty of potentially disastrous architecture. The collages depict expansive Texas highways set against allegorical landscapes composed of world atlas maps. The oblique aerial images of these Dallas-Fort Worth highway interchanges are based on photographs taken by Cusick from a helicopter. Having recently been reconstructed to accommodate massive increases in traffic along the NAFTA superhighway corridor, Cusick interprets these highway interchanges as portals into history, erected upon buffalo migration paths, tribal hunting grounds, trading posts and railroad depots. To create these imaginary landscapes, maps of individual countries were cut out along their natural contours and political boundaries and then reconfigured into an assemblage of geology gone awry. Cusick has drawn from the ravaged topography of abandoned open pit and strip mines, wishing to capture the caustic quality of these chemical wastelands. Color choices are poignant and disturbing; such as the cyan blue that poisons the coastline in Chasing the Dragon (2006). Maps are still masterfully cut and inlaid, but some areas are left untreated while others appear to have already eroded. These tiny deviations in craftsmanship are for Cusick apocalyptic gestures, revealing the ominous end to our civil ambitions.

136 DESIGN Matthew Cusick COUNTRY/REGION The USA

Chasing the Dragon, 2006 Inlaid maps and acrylic on wood panel 40 x 64 inches

Marc Rivers, 2009 Inked maps and acrylic on wood panel 48 x 77 inches

Course of Empire (Mixmaster II), 2006 Inlaid maps and acrylic on wood panel 48 x 77 inches

MAPPING THE PHYSICAL ENVIRONMENT

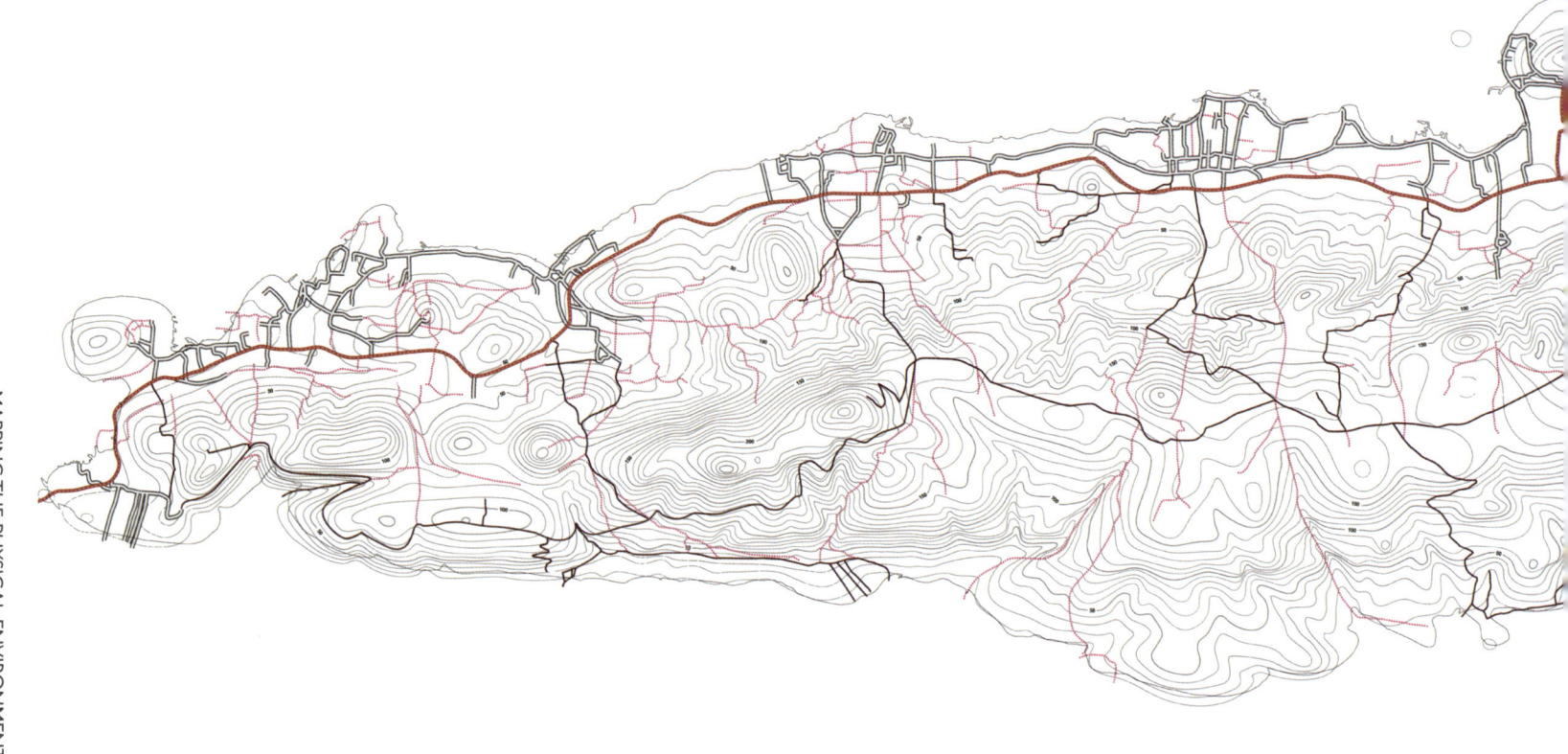

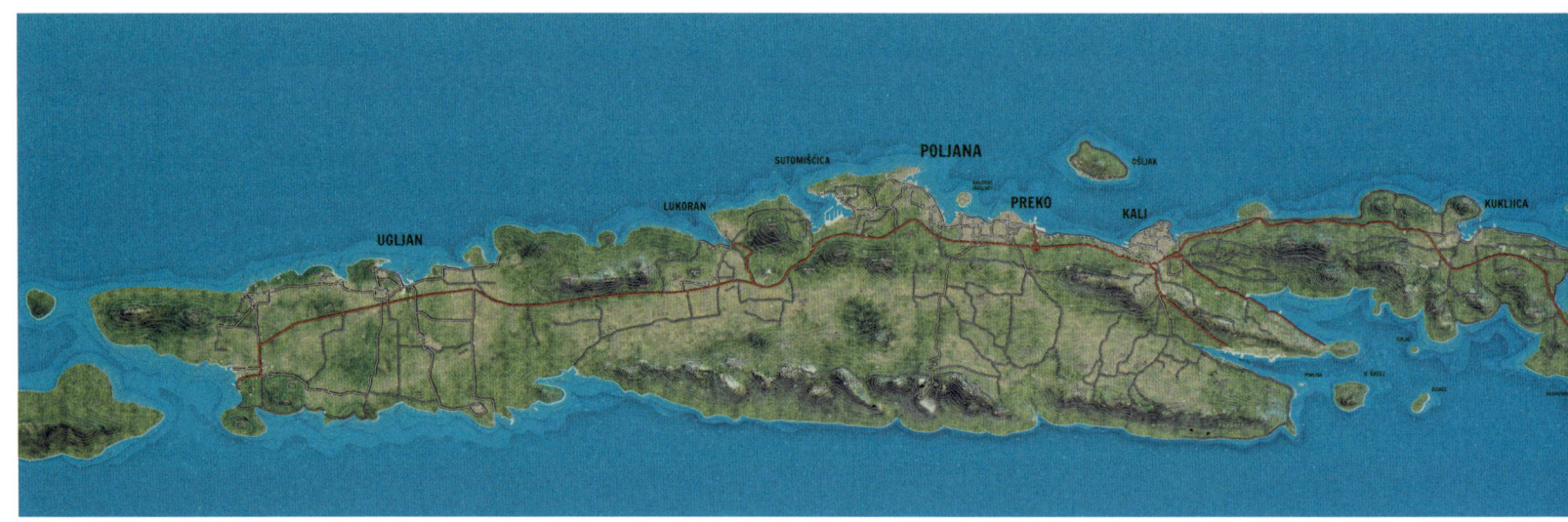

Ugljan & Pasman

This map is a combination of vector and raster graphics. The work shows the map of the two beautiful islands of Ugljan and Pasman located in the Zadar Archipelago. It was created with concentration, patience, and eye for detail. The map has twenty separate thematic layers and can be modified to a desired size without loss of any resolution.

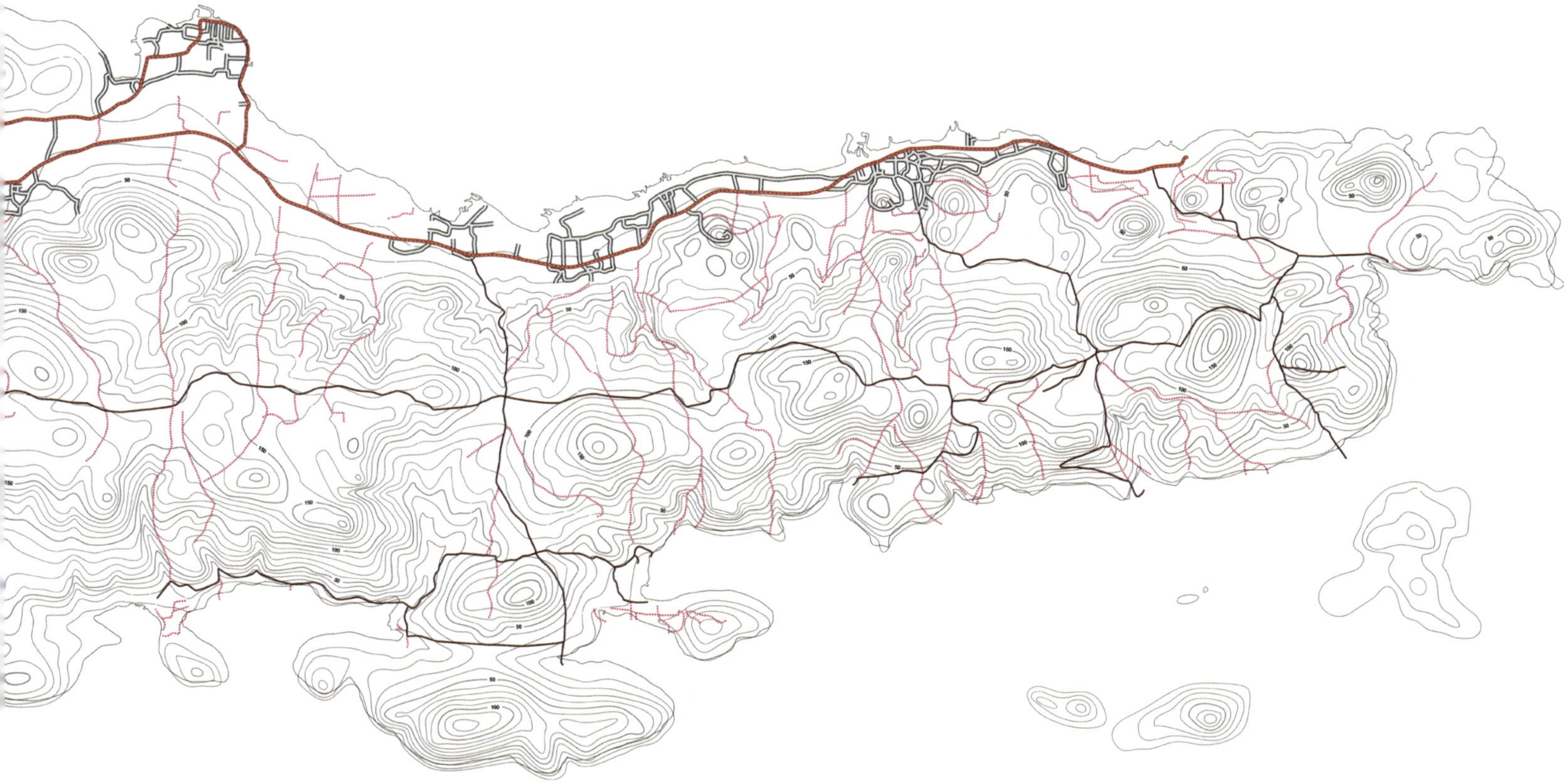

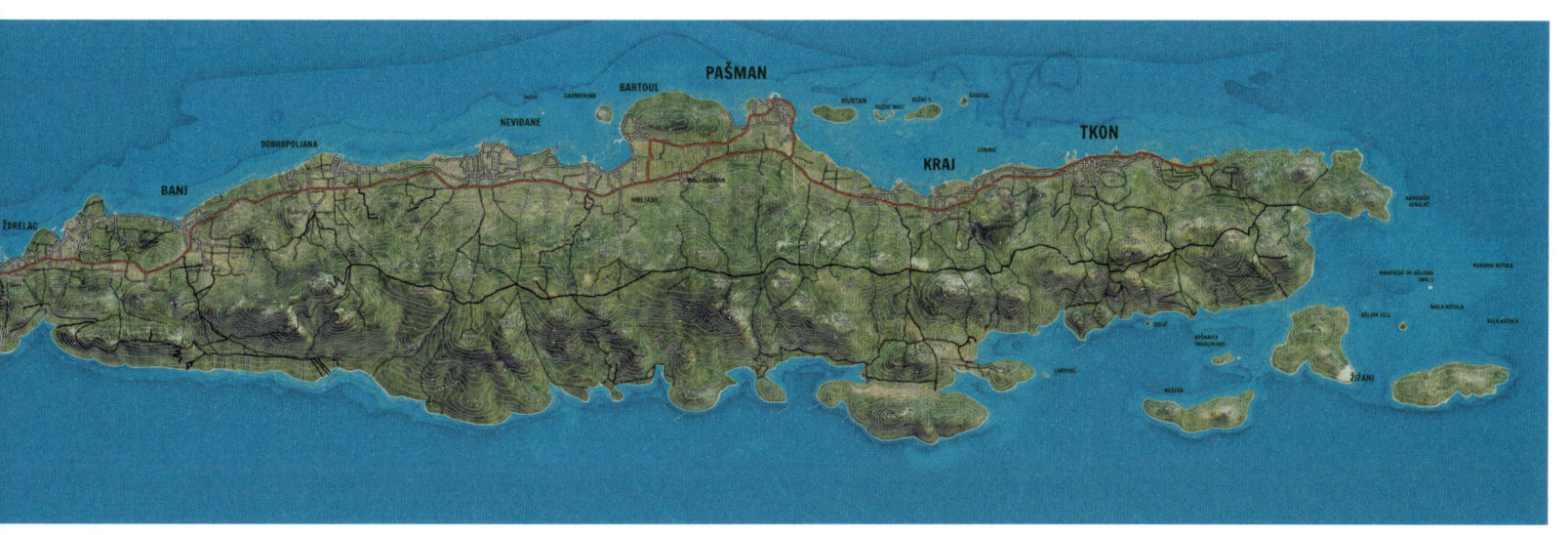

DESIGN Martina Sikiric COUNTRY/REGION Croatia 143

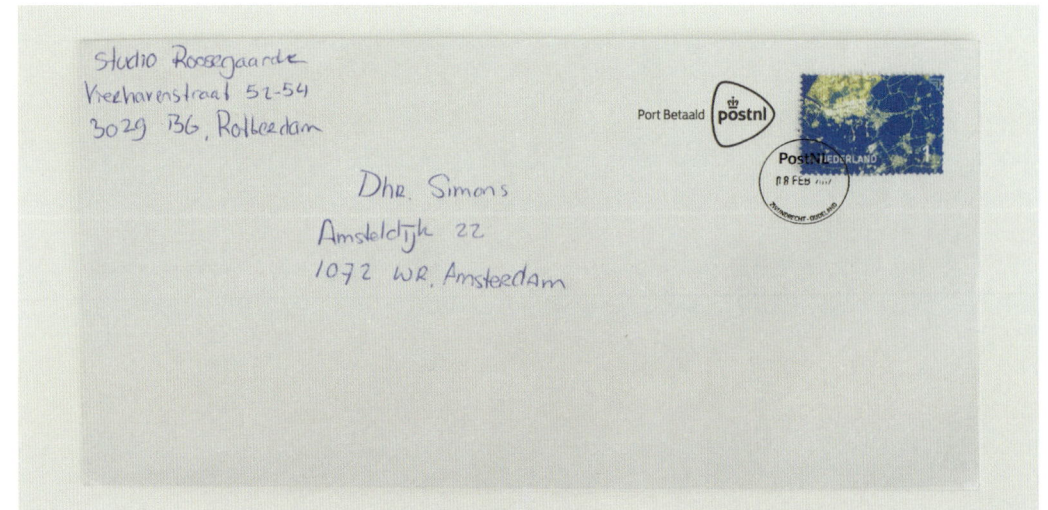

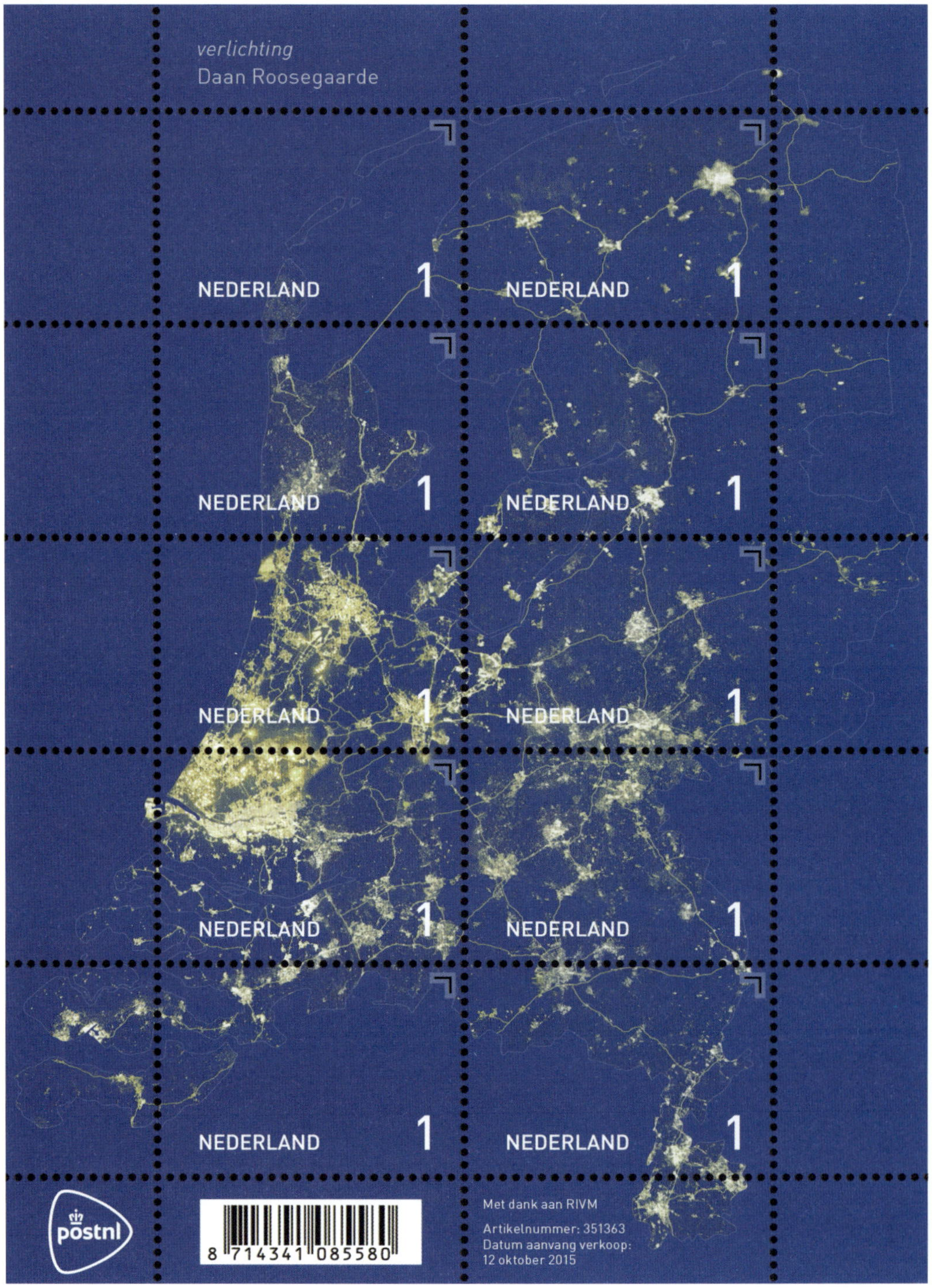

Enlightened Postal Stamps

This project is a map as stamps. Each separate stamp features a different piece of the Netherlands and together the stamps form a complete map when joined in the correct order. When designing the stamps, Roosegaarde was inspired by aerial photos taken by André Kuipers and satellite photos of the Netherlands, after which he incorporated his own interpretation, showing the Netherlands in a unique way—as a network of light seen from space, shaped by the cities and roads of the country.

The result is a Netherlands map we have never seen before. The amazing network of light establishes connections, making an ominous and poetic piece. It gives us an insight into our behavior and the impact it has on our landscape. In this way, the map as stamps offers both literal and figurative enlightenment.

DESIGN Daan Roosegaarde   COUNTRY/REGION The Netherlands

Nakhon Si Thammarat Province

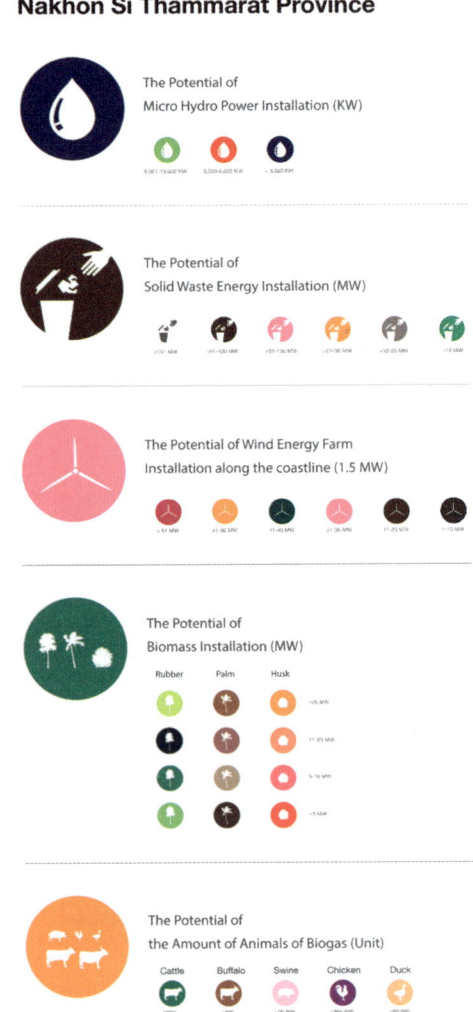

The Potential of Micro Hydro Power Installation (KW)

The Potential of Solid Waste Energy Installation (MW)

The Potential of Wind Energy Farm Installation along the coastline (1.5 MW)

The Potential of Biomass Installation (MW)

The Potential of the Amount of Animals of Biogas (Unit)

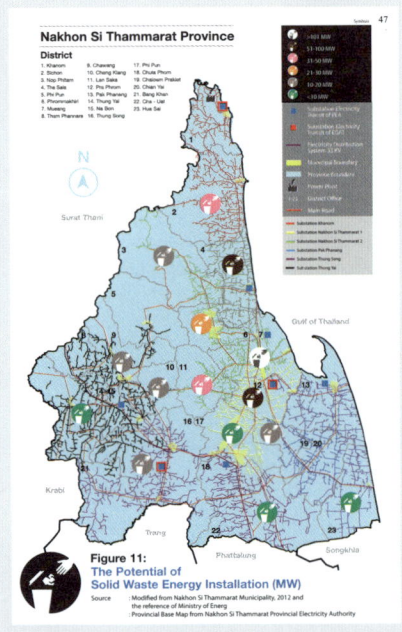
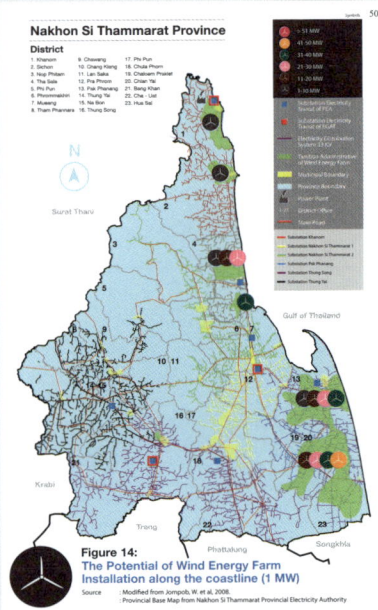
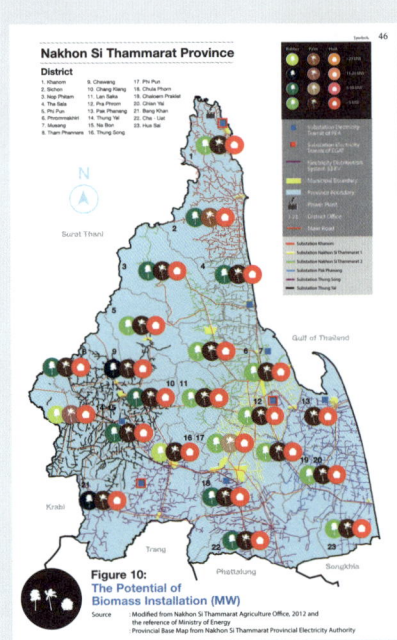
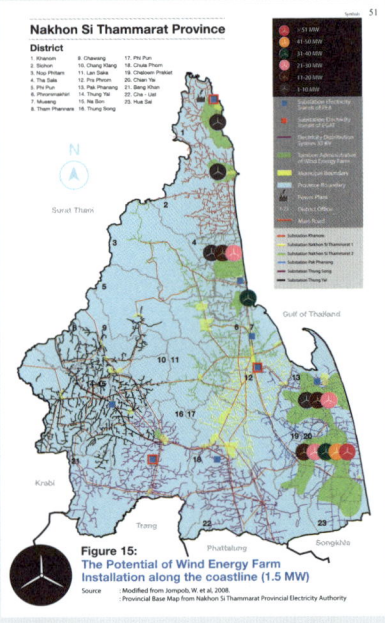
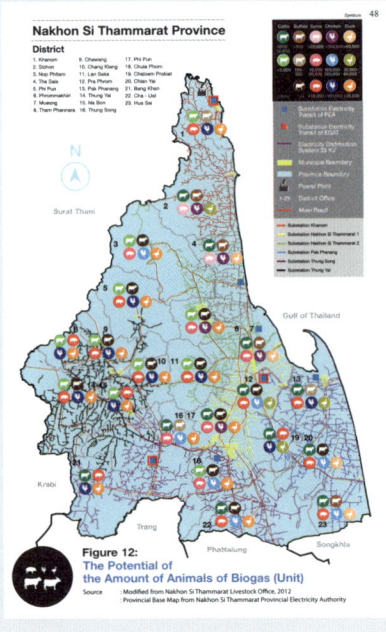
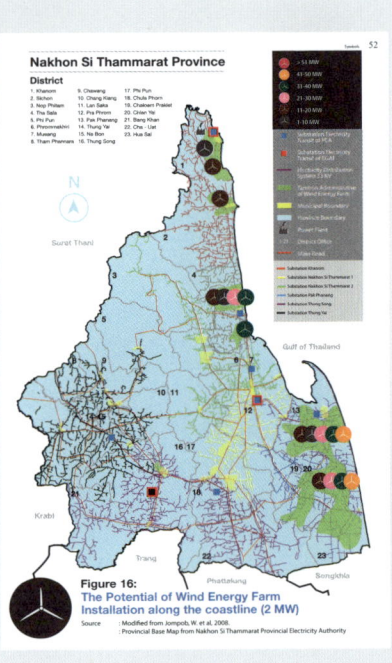

Nakorn Si Thammarat Power Development Plan

This map visualizes the power development plan for Nakorn Si Tammarat, a city in southern Thailand. All reused bio power comes from wind, livestock, plants, waste, and water.

DESIGN   MYDM co., ltd   COUNTRY/REGION   Thailand

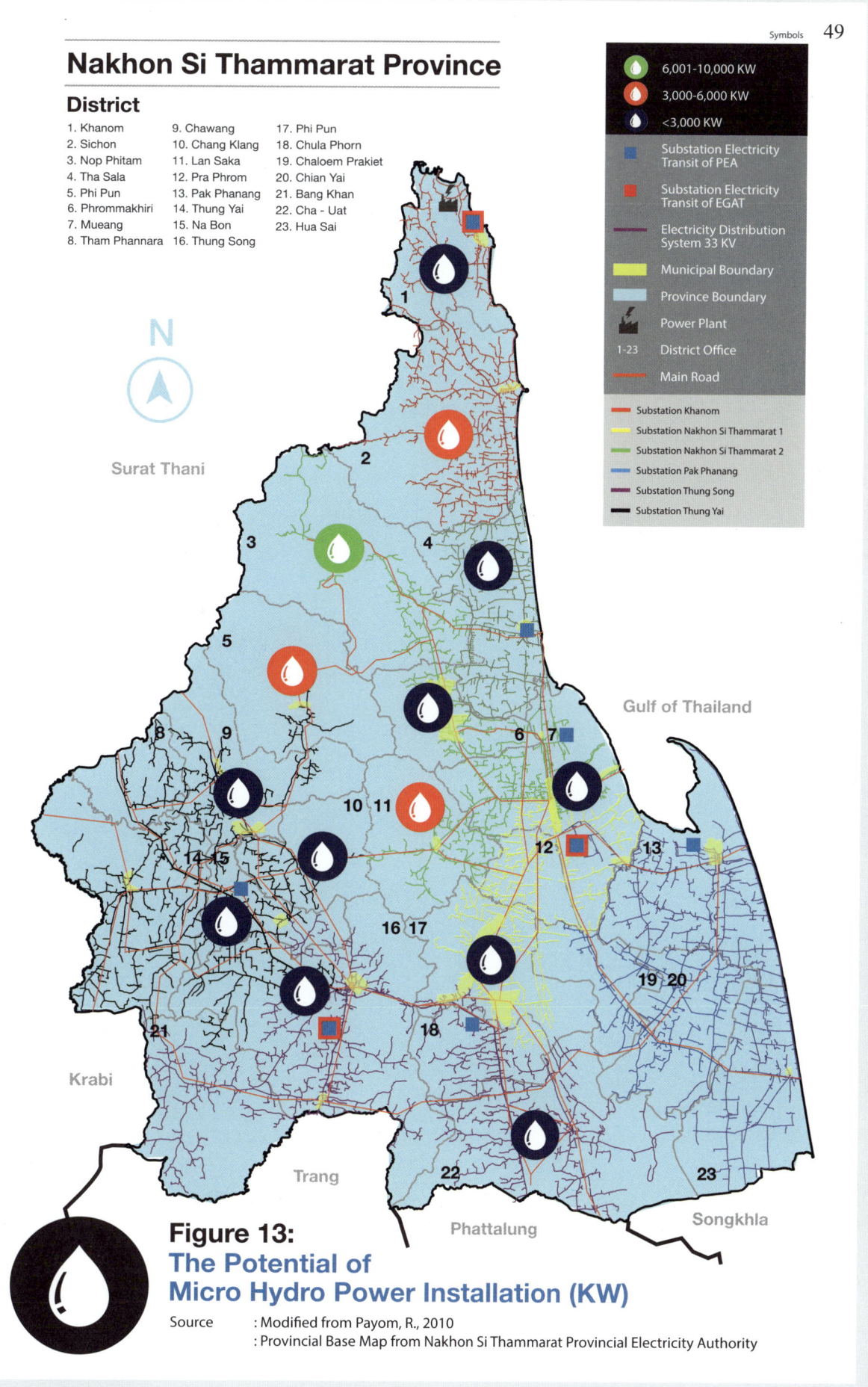

Figure 13:
The Potential of Micro Hydro Power Installation (KW)

Source : Modified from Payom, R., 2010
: Provincial Base Map from Nakhon Si Thammarat Provincial Electricity Authority

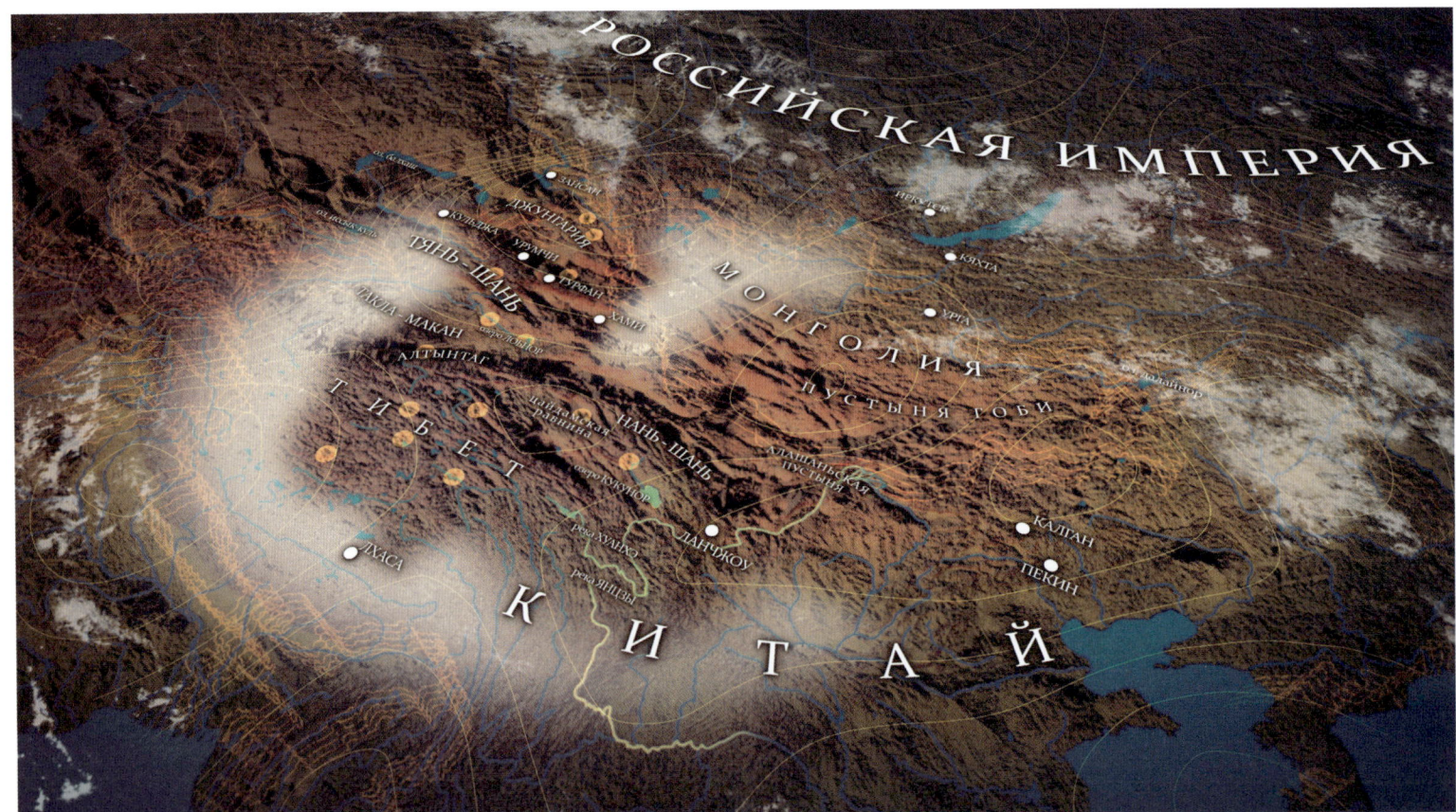

Nikolay Prjevasky

This map illustrates the adventures and discoveries of a Russian geographer and explorer Nikolai Przewalski. It is part of a documentary about the explorer. The map was made in a similar style to motion graphics. It puts visual emphasis on the places on the path which Nikolai Przewalski travelled through to his never-reached ultimate goal—the Tibetan city of Lhasa.

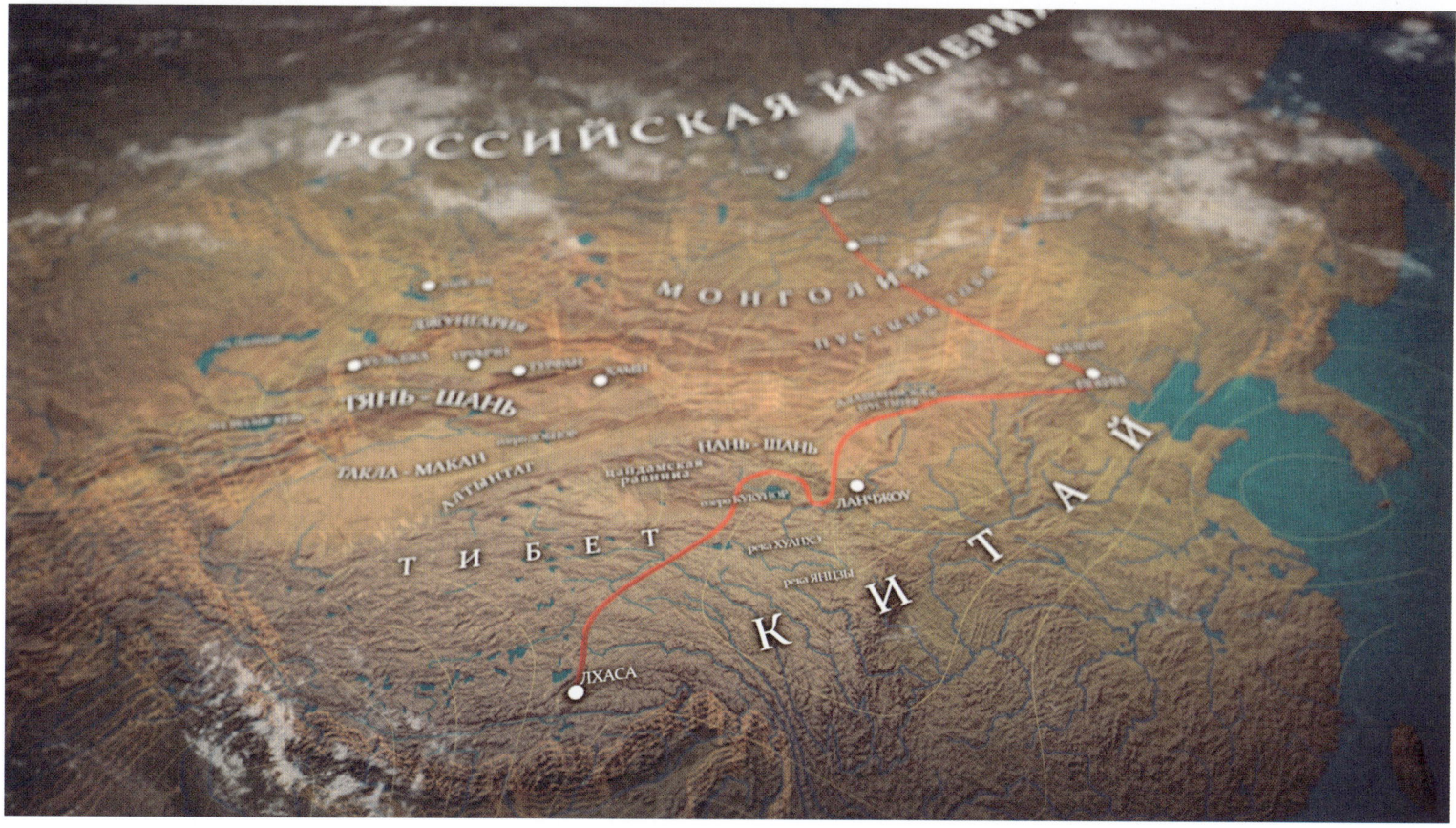

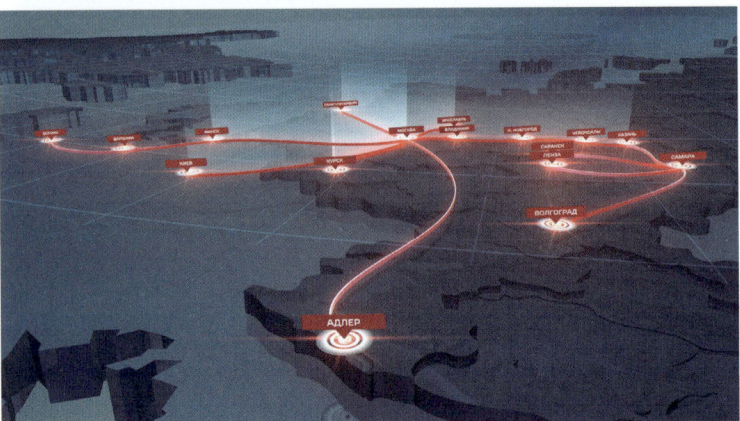

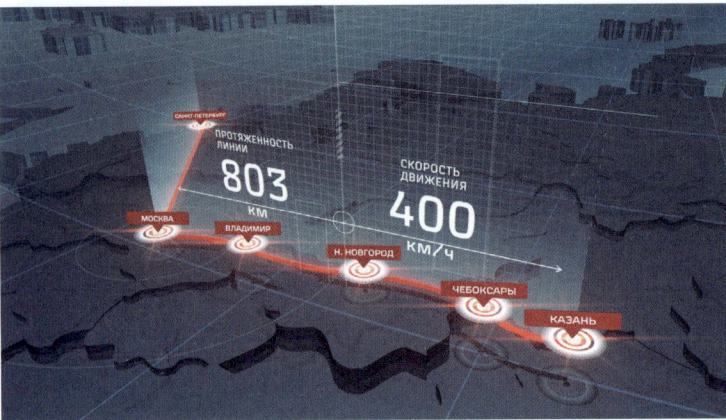
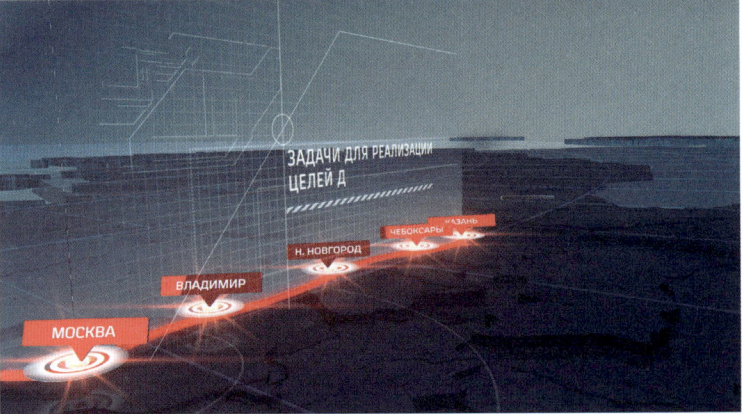

RZD Map

This interactive map is to show the prospect of rail transport development in Russia, which is planning a high-speed rail line that will link remote regions of the country into a single network. To achieve the objective required a bright and contemporary infographic work, showing the relevant map and geo-information. The work was featured as part of a presentation to a major railway company and is a motion graphic presentation made by means of the program Cinema 4d.

DESIGN Dmitriy Vorontzov   COUNTRY/REGION Russia

Two Maps for "Summer Day Tripping in the UK"

Rowland was commissioned by Sainsbury's Bank Money Matters Magazine to create two maps for their summer issues. The subject was day trips in the UK—one aimed at readers interested in family entertainment and one of general interest. Each place name was individually hand lettered and this lettering was also used to punctuate the body copy of the articles.

Topographic Typography

Inzon represents each province of the Philippines by their names in this map. But some of the names don't quite fit the shape of each border; the obscurity creates a challenge for viewers familiar with Filipino topography to identify each province by its name. It also serves as a convenient reference.

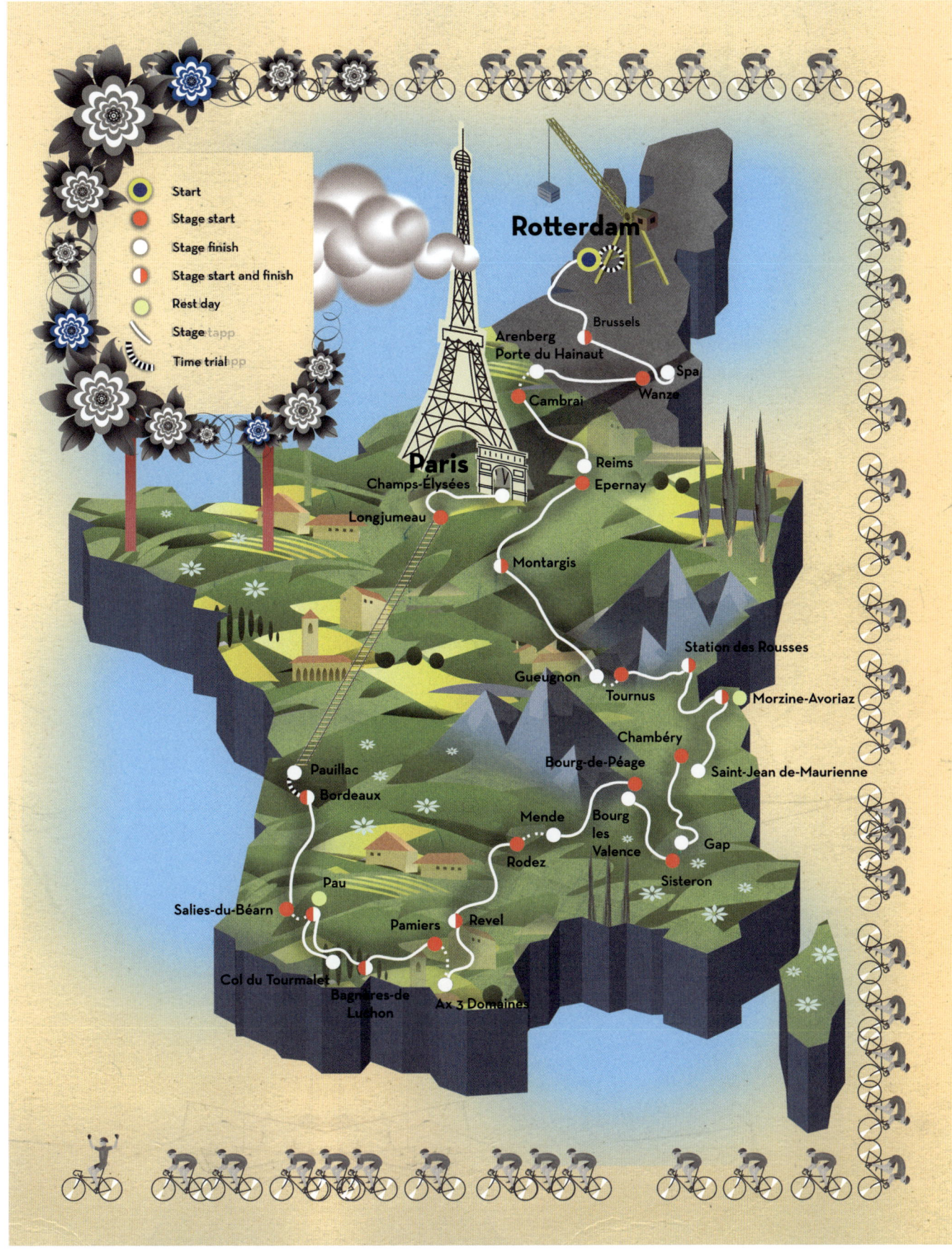

Tour de France Map

This map was for Kadens magazine, a Swedish cycling magazine. It shows Tour de France 2010, one of cycling's Grand Tours. The race visited three countries: the Netherlands, Belgium and France, and finished in Paris.

DESIGN Nils-Petter Ekwall   COUNTRY/REGION Sweden

Ad Summit 2014 Subic Map

This project is a partnership between Jo Malinis and Raxenne Maniquiz. It is a map of Subic for Adobo Magazine's Ad Summit Pilipinas 2014 Festival Guide.
The map on top is a zoomed-in version of the bay area. The other half shows the whole of Subic.

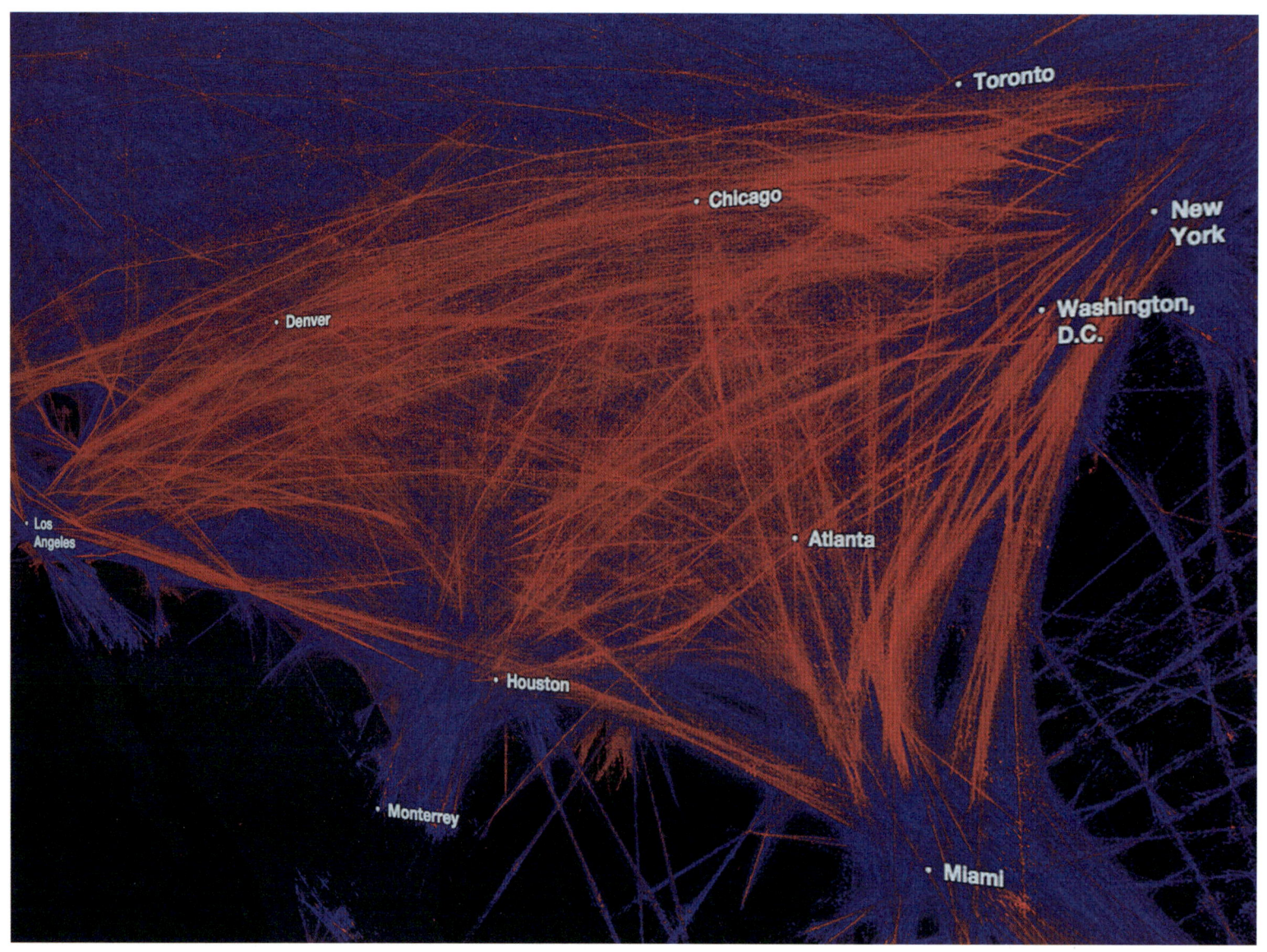

Flight Paths of the World Map

This map aimed to visualize plane routes around the world. The map drawing was based on the flight data collected during the whole month of October in 2012. The results of such monitoring amounted to about one billion "dots" which then were put on a map. Using red for lower altitude flights and blue for higher, the map produces some very beautiful patterns of different flight paths across the world. It also allows visualization of the "roads", "highways", "intersections" and "junctions" used by airplanes.

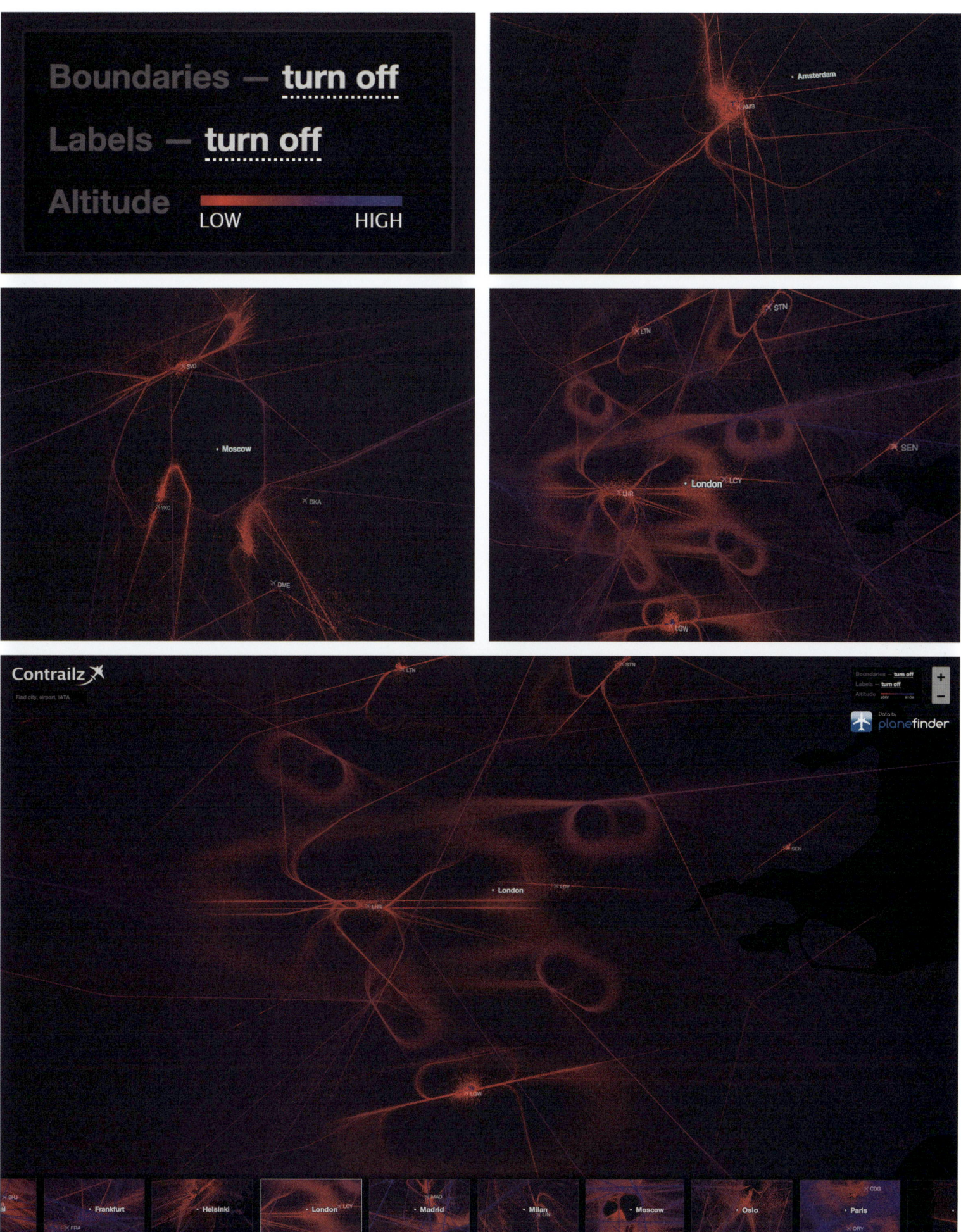

DESIGN Alexey Papulovskiy & Nikolay Guryanov COUNTRY/REGION Russia

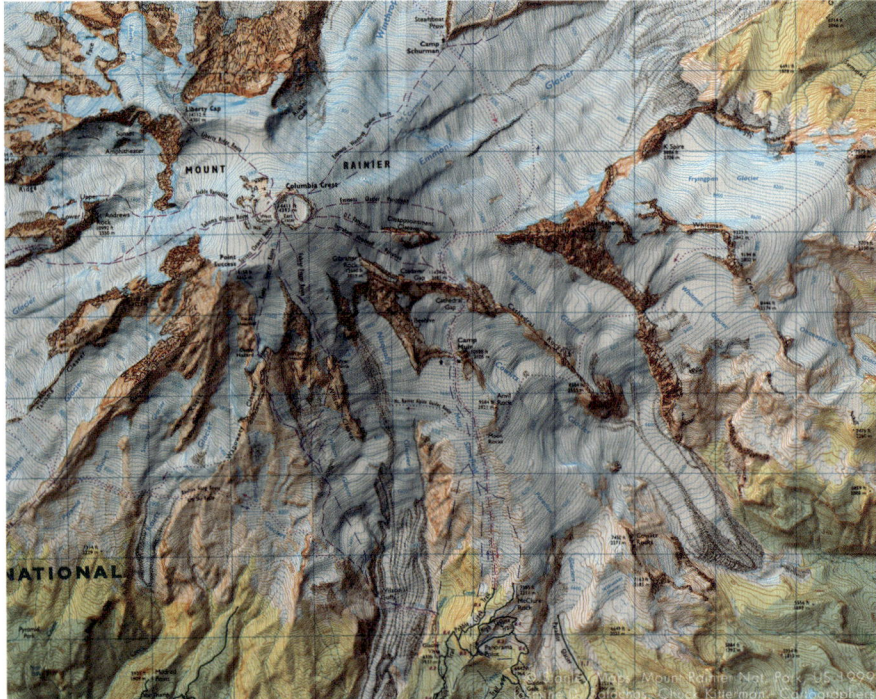

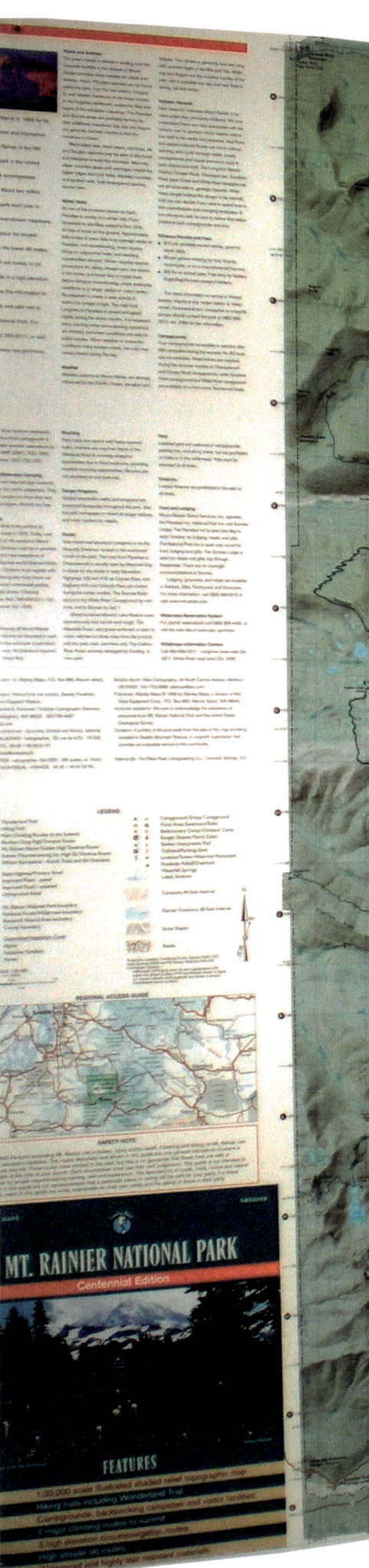

Mount Rainier National Park Map

Mount Rainier National Park Centennial Edition is a huge elaborately colored relief map of Mount Rainier National Park published by Stanley Friedman Maps Company for the park's centennial (1999). The map was drawn according to the standards of National Geographic Institute (France) topographical maps.

TERRAIN FEATURES   Jasmine Desclaux-Salachas   COUNTRY/REGION   France

© Stanley Maps, Mount Rainier Nat. Park, US 1999
Jasmine D. Salachas, Chuck Kitterman - Cartographers

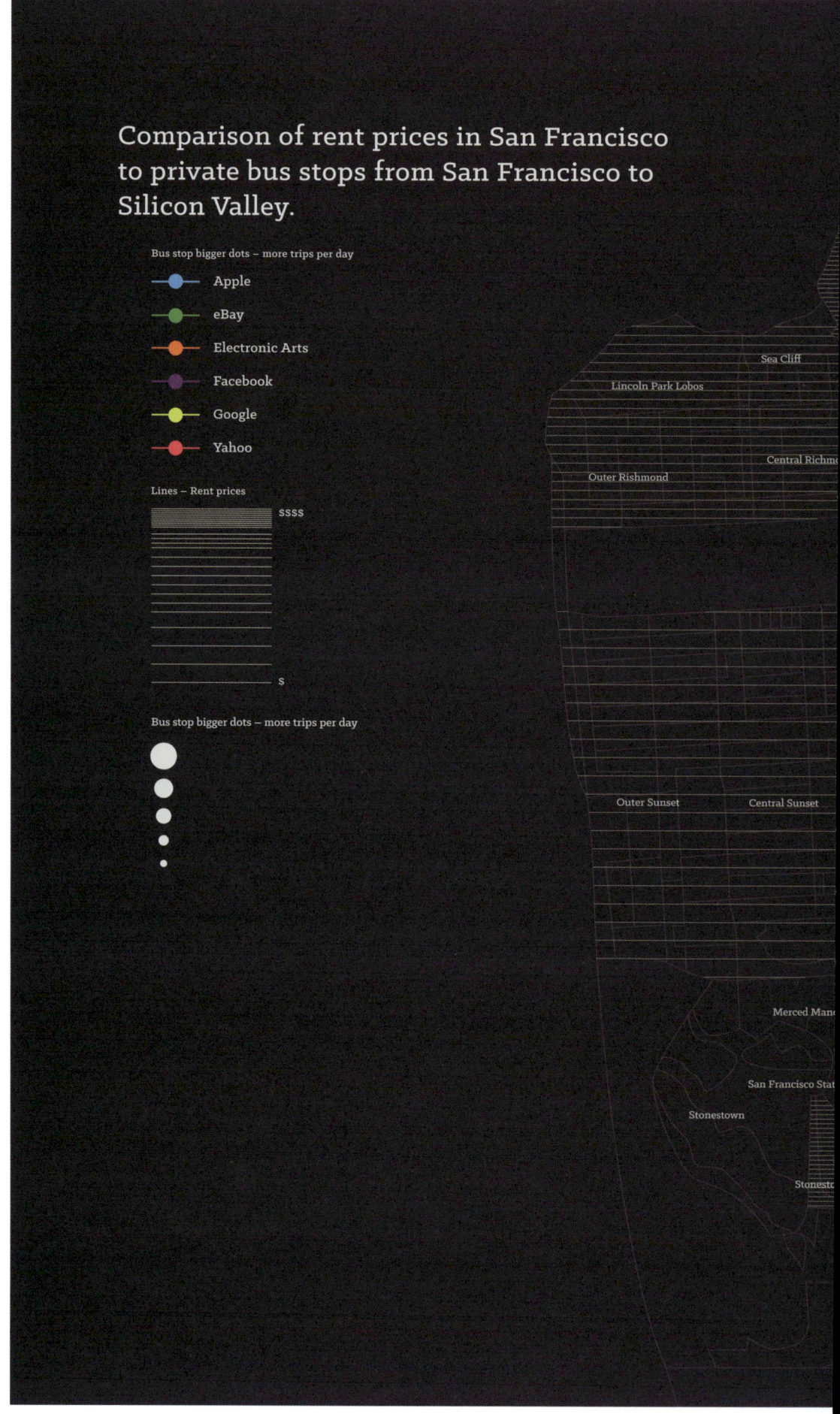

Mapping San Francisco

This infographic map compares rent prices in San Francisco to private bus stops from San Francisco to Silicon Valley. By combining these two things that San Francisco can't stop thinking about into a really pretty map infographic, the map matter-of-factly shows the correlation of rent prices and proximity to private bus stops for shuttles to Silicon Valley. A few colors and different spacing and lines were adopted to create a simple yet clear and direct map.

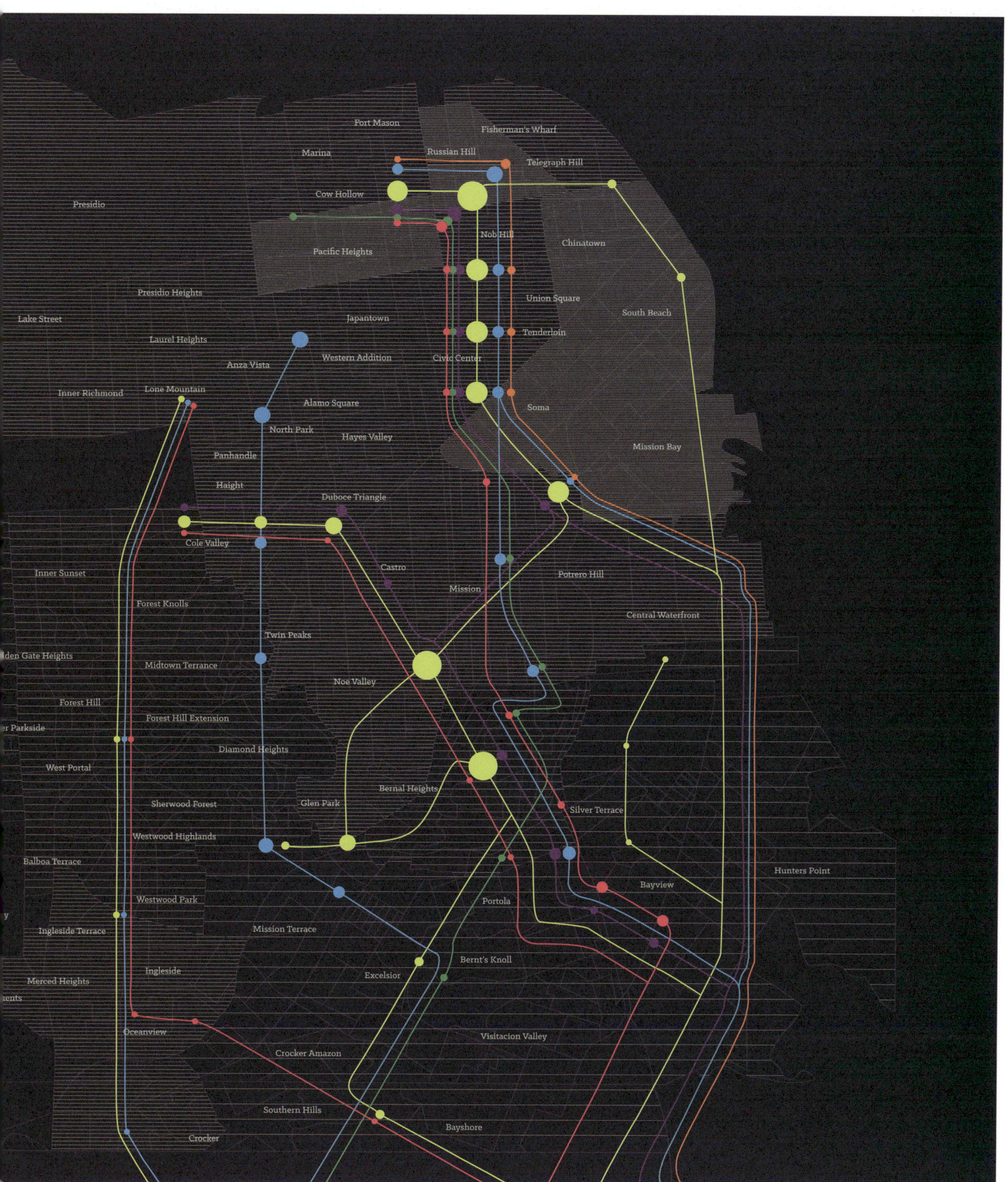

DESIGN Suwanna Ruayrinsaowarot   COUNTRY/REGION The USA   163

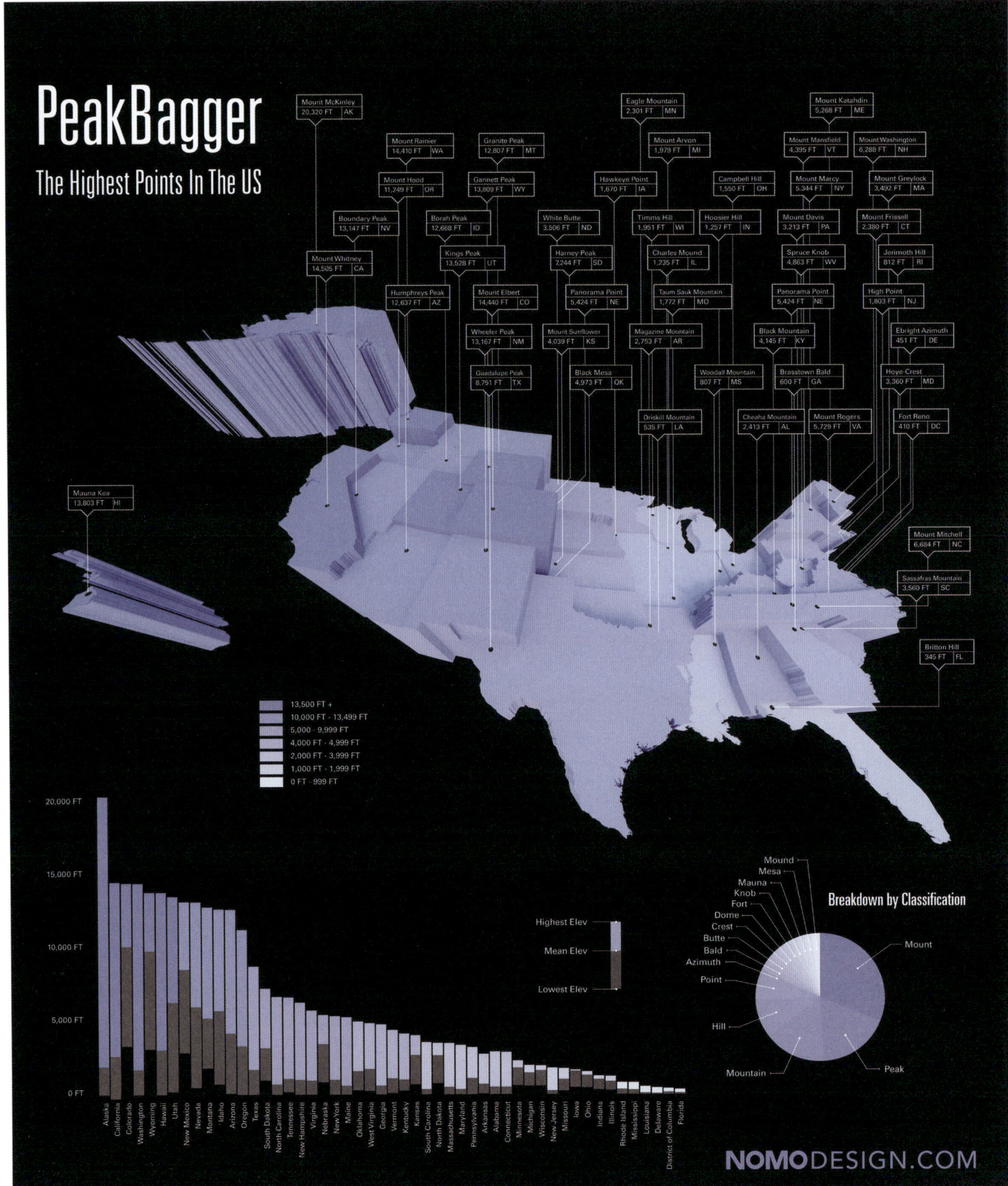

PeakBagger: The Highest Points in the US

PeakBagger is a 3D visualization of the highest points in each state of America. A pie chart illustrates the classification name of each landmark—mountains, peaks and points, etc. A bar chart also illustrates the highest, average, and lowest elevations for each state.

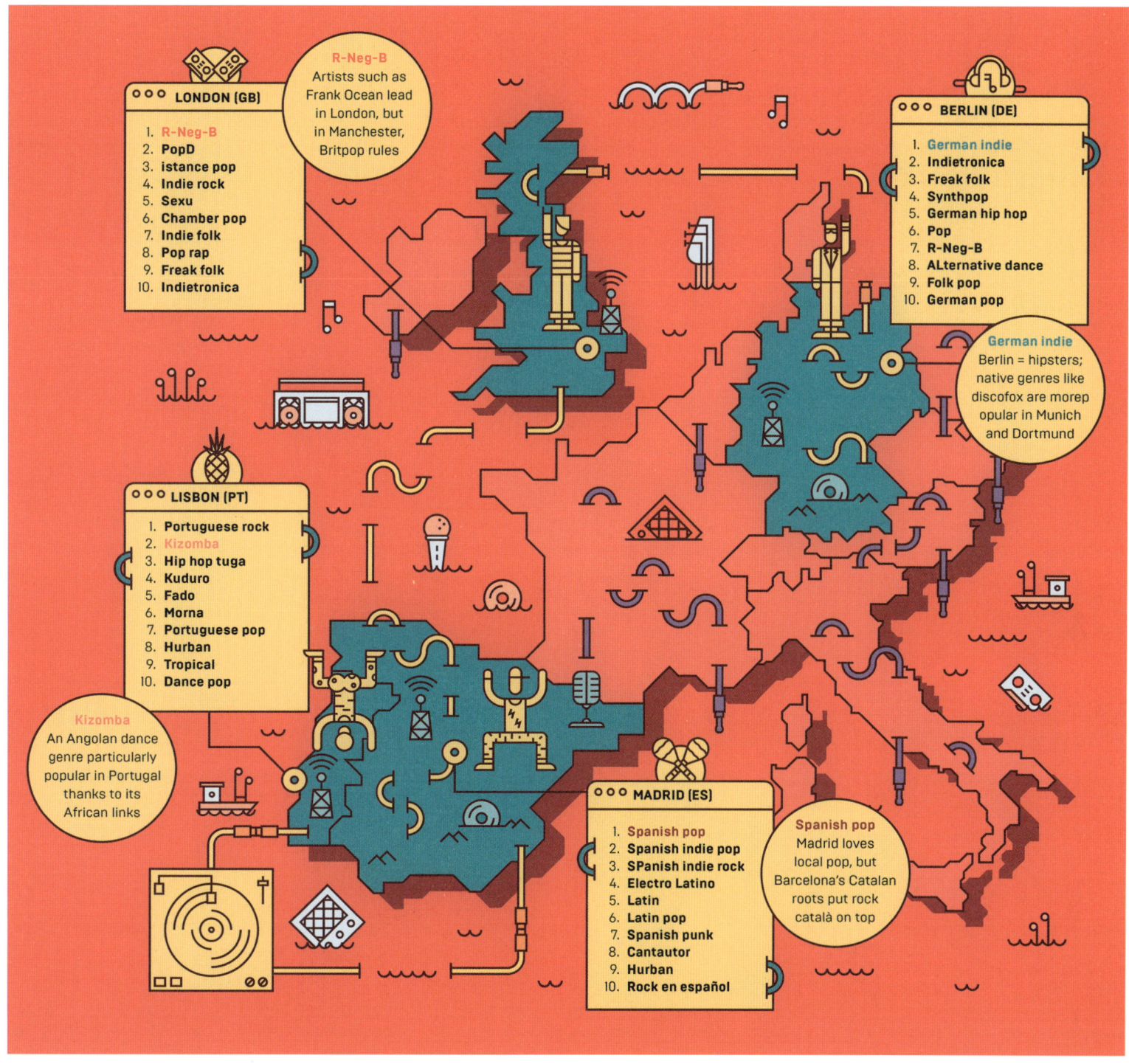

Spotify EU Map

This map was made to find out what EU is listening to. The design duo represents the different moods of four countries with a sparkling, colorful map of Middle West Europe. Light blue countries are the ones playing music, while the light red ones do the opposite. A compass is added as a record player wired to all the active countries on the map: UK, Germany, Portugal and Spain. Small characters represent the main music style played in each country. Everything is smartly integrated to make a simple and impactful composition.

DESIGN   relajaelcoco studio   COUNTRY/REGION   Spain

Ougeoma

Ougeoma is a fictional map created purely for the joy of world making. The work is rendered richly as if it were a physical landscape in topographic view rather than a graphical, informative map.

DESIGN Sam Williams Studio   COUNTRY/REGION Australia

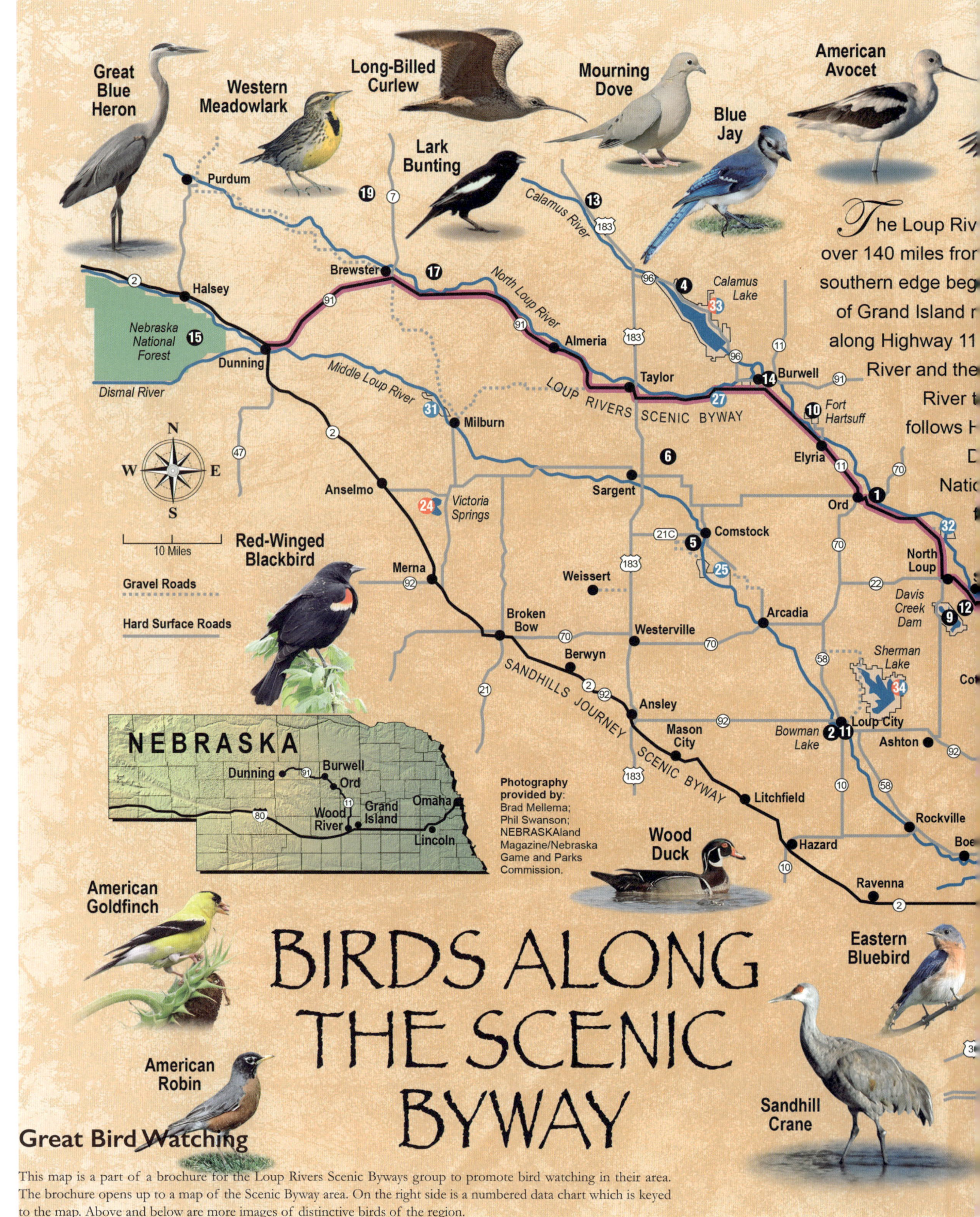

This map is a part of a brochure for the Loup Rivers Scenic Byways group to promote bird watching in their area. The brochure opens up to a map of the Scenic Byway area. On the right side is a numbered data chart which is keyed to the map. Above and below are more images of distinctive birds of the region.

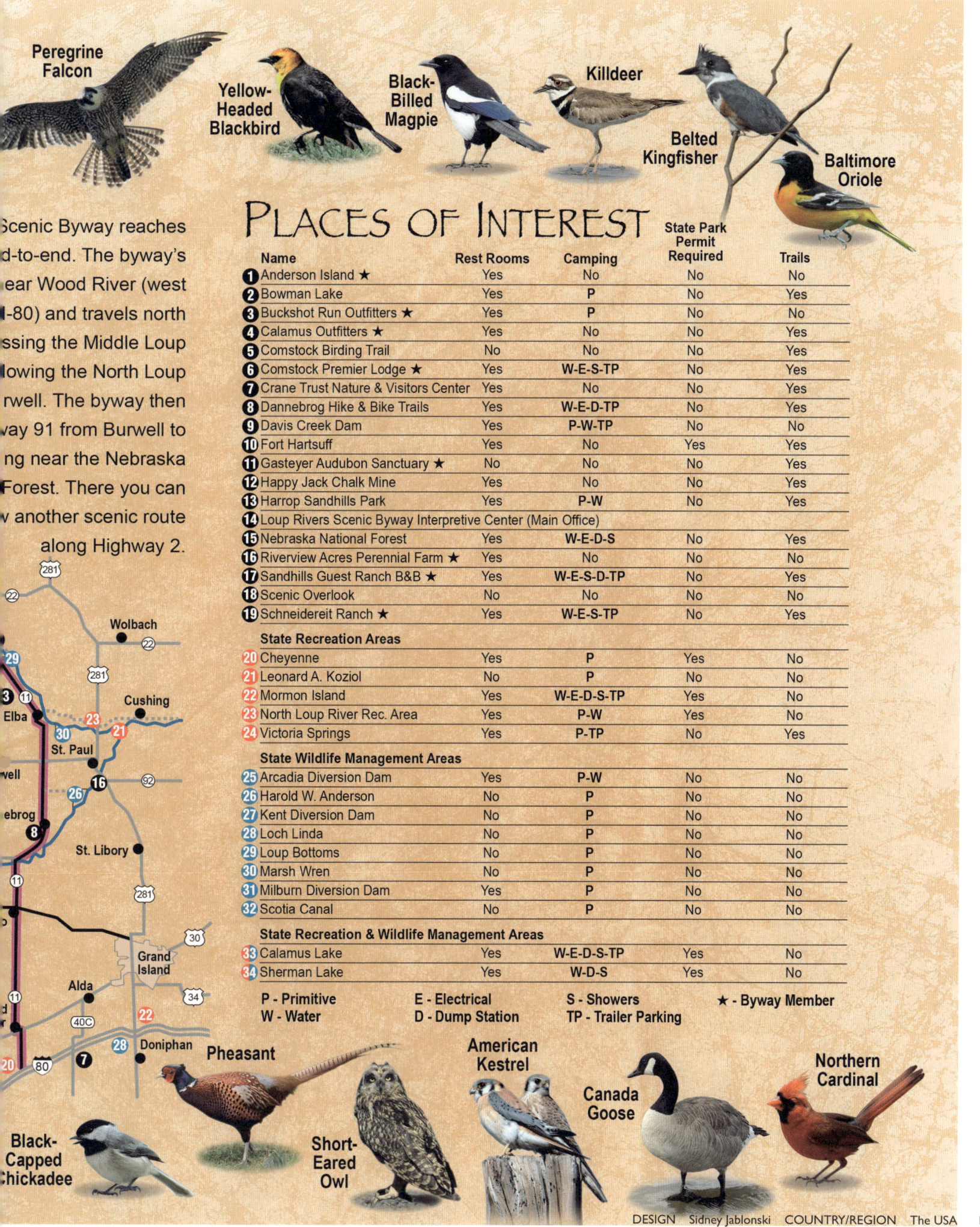

Places of Interest

	Name	Rest Rooms	Camping	State Park Permit Required	Trails
1	Anderson Island ★	Yes	No	No	No
2	Bowman Lake	Yes	P	No	Yes
3	Buckshot Run Outfitters ★	Yes	P	No	No
4	Calamus Outfitters ★	Yes	No	No	Yes
5	Comstock Birding Trail	No	No	No	Yes
6	Comstock Premier Lodge ★	Yes	W-E-S-TP	No	Yes
7	Crane Trust Nature & Visitors Center	Yes	No	No	Yes
8	Dannebrog Hike & Bike Trails	Yes	W-E-D-TP	No	Yes
9	Davis Creek Dam	Yes	P-W-TP	No	No
10	Fort Hartsuff	Yes	No	Yes	Yes
11	Gasteyer Audubon Sanctuary ★	No	No	No	Yes
12	Happy Jack Chalk Mine	Yes	No	No	Yes
13	Harrop Sandhills Park	Yes	P-W	No	Yes
14	Loup Rivers Scenic Byway Interpretive Center (Main Office)				
15	Nebraska National Forest	Yes	W-E-D-S	No	Yes
16	Riverview Acres Perennial Farm ★	Yes	No	No	No
17	Sandhills Guest Ranch B&B ★	Yes	W-E-S-D-TP	No	Yes
18	Scenic Overlook	No	No	No	No
19	Schneidereit Ranch ★	Yes	W-E-S-TP	No	Yes
	State Recreation Areas				
20	Cheyenne	Yes	P	Yes	No
21	Leonard A. Koziol	No	P	No	No
22	Mormon Island	Yes	W-E-D-S-TP	Yes	No
23	North Loup River Rec. Area	Yes	P-W	Yes	No
24	Victoria Springs	Yes	P-TP	No	Yes
	State Wildlife Management Areas				
25	Arcadia Diversion Dam	Yes	P-W	No	No
26	Harold W. Anderson	No	P	No	No
27	Kent Diversion Dam	No	P	No	No
28	Loch Linda	No	P	No	No
29	Loup Bottoms	No	P	No	No
30	Marsh Wren	No	P	No	No
31	Milburn Diversion Dam	Yes	P	No	No
32	Scotia Canal	No	P	No	No
	State Recreation & Wildlife Management Areas				
33	Calamus Lake	Yes	W-E-D-S-TP	Yes	No
34	Sherman Lake	Yes	W-D-S	Yes	No

P - Primitive   E - Electrical   S - Showers   ★ - Byway Member
W - Water   D - Dump Station   TP - Trailer Parking

DESIGN Sidney Jablonski   COUNTRY/REGION The USA

MAPPING THE PHYSICAL ENVIRONMENT

Seoul Metro

With emphasis on vibrancy, balance, and consistency, the inspiration for this metro map came from the Asian philosophical concept of Yin and Yang which can be found in many traditional Korean designs (including the national flag). The Yin-Yang element is found consistently throughout the design such as the river in the center and transfer stations.

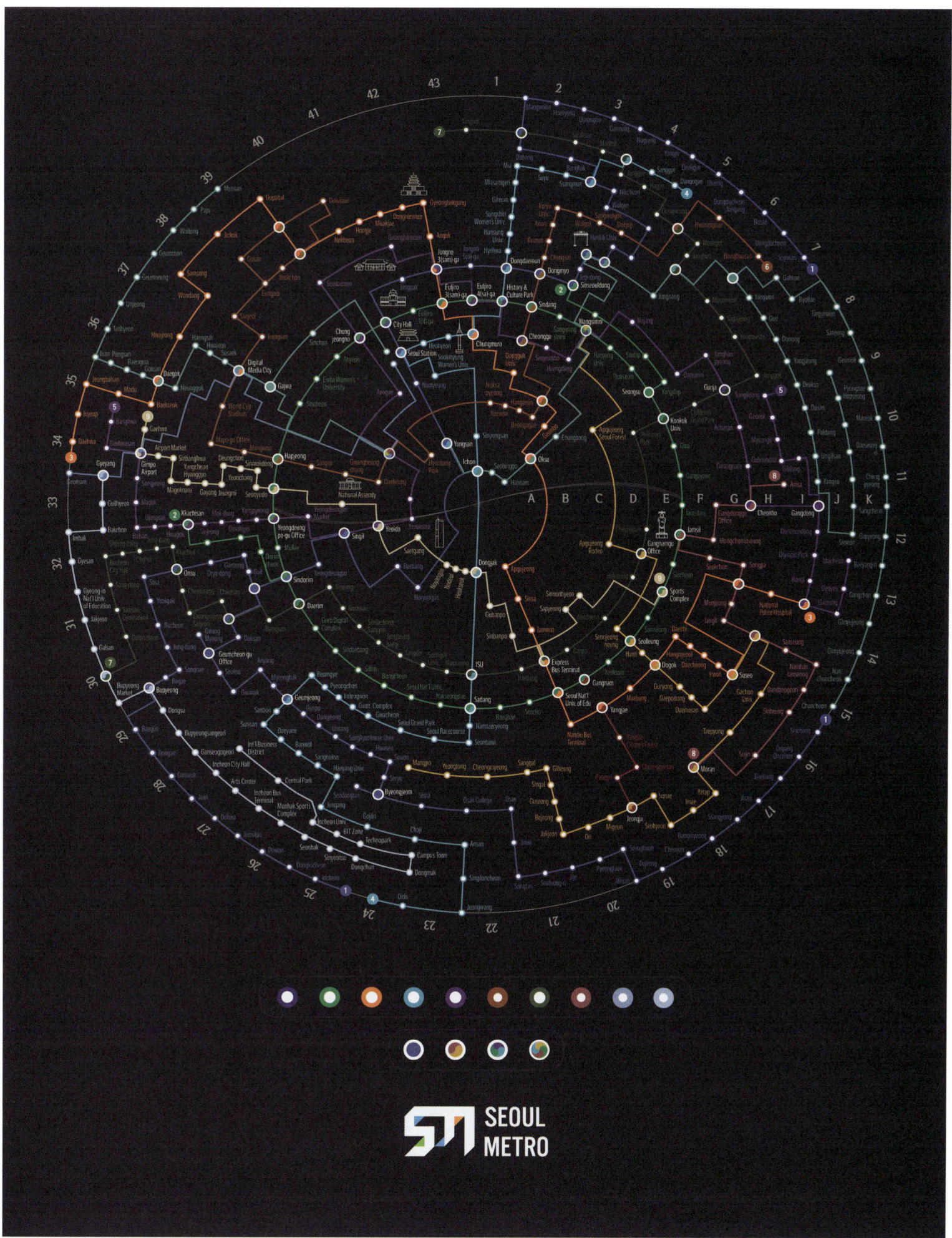

Flowing City Map

These maps are part of a series of maps entitled Flowing City Map created for an exhibition which explores cities, their environments and the relation between them. Chaotic Atmospheres represented the cities' influence on the environment as a kind of invisible fluid that overflows from the city to its surroundings. City maps were retrieved and roads were highlighted. Then, with the help of a 3D processing software, the artist created the "fluid" by simulating an erosion on the maps.

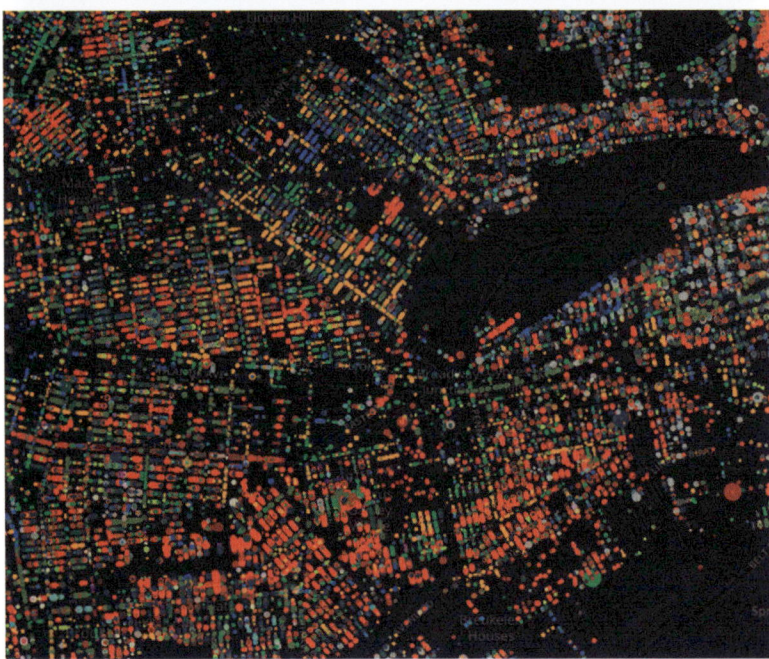

- ☐ All Trees
- ■ American Linden
- ■ Amur Maple
- ■ Baldcypress
- ■ Birch
- ■ Black Locust
- ■ Callery Pear
- ■ Chinese Elm
- ■ Crabapple
- ■ Dawn Redwood
- ■ Eastern Redbud
- ■ Eastern White Pine
- ■ English Oak
- ■ European Hornbeam
- ■ Flowering Cherry
- ■ Flowering Dogwood
- ■ Ginkgo
- ■ Goldenrain Tree
- ■ Green Ash
- ■ Hackberry
- ■ Hawthorn
- ■ Hedge Maple
- ■ Honeylocust
- ■ Horsechestnut
- ■ Japanese Maple
- ■ Japanese Pagoda Tree
- ■ Japanese Tree Lilac
- ■ Japanese Zelkova
- ■ Katsura Tree
- ■ Kentucky Coffeetree
- ■ Littleleaf Linden
- ■ London Planetree
- ■ Magnolia
- ■ Mulberry
- ■ Northern Red Oak
- ■ Norway Maple
- ■ Pin Oak
- ■ Poplar
- ■ Purpleleaf Plum
- ■ Red Maple
- ■ Sawtooth Oak
- ■ Schubert Cherry
- ■ Serviceberry
- ■ Silver Linden
- ■ Silver Maple
- ■ Sugar Maple
- ■ Swamp White Oak
- ■ Sweetgum
- ■ Sycamore Maple
- ■ Tree of Heaven
- ■ Tulip Tree
- ■ White Oak
- ■ Willow Oak

FILTER BY SPECIES ▲

NYC Street Trees by Species

New York City's urban forest provides numerous environmental and social benefits, and street trees compose roughly one quarter of that canopy. The NYC Street Trees by Speices map shows the distribution and biodiversity of the city's street trees based on the 2005 New York City tree census, conducted by the Department of Parks and Recreation.

Long Island City, Queens, New York

MAPPING THE PHYSICAL ENVIRONMENT

Environs Map Series

The Environs Map Series is a living document of psycho-geographic maps. They are comprised of original photography, image compositing, and illustration collaged into an exact geography of a place. They are at once highly objective and subjective. The maps seek to capture the essence of the environment at the moment that the information and imagery are captured and created.

Carroll Gardens, Brooklyn, New York

178　Williamsburg, Brooklyn, New York

Chapel Hill, North Carolina

Mappemonde

A world map hammered in the wall at the scale of an embrace reveals continents and nations in the gaps of the surface, holes and cracks for land, smooth and white surfaces for oceans.
Here, the world map is not only a picture of the world, but reminds the viewer of the physical presence of the artist. Its scale put in front of the exhibition wall, the distance between his body and the piece of work, and the space between his two arms defines the surface of the piece. Just imagine him engraving the World's picture by heart, hammer in hand, when confronted with a blank wall.

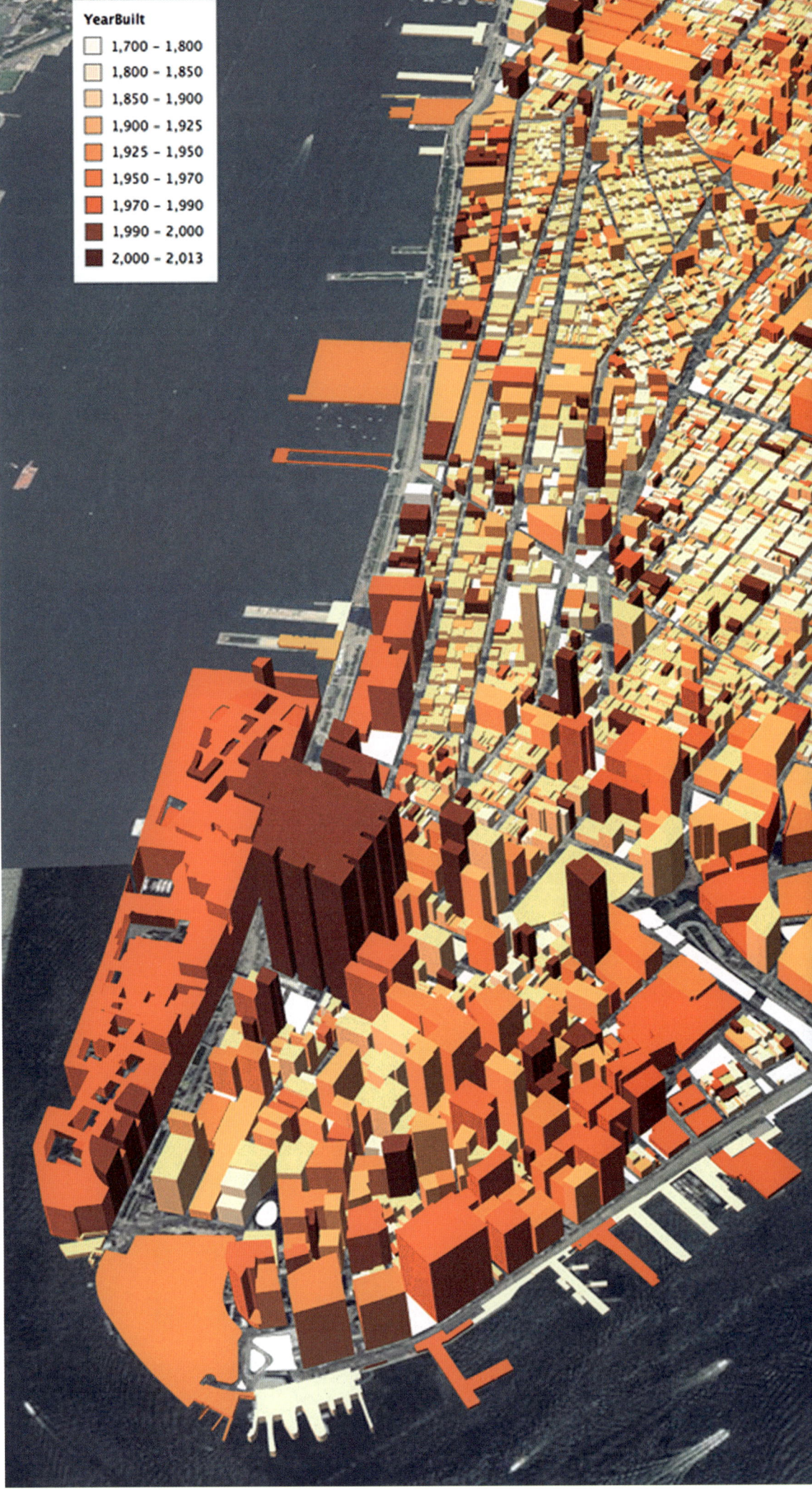

Lower Manhattan by Year Built

The 3D rendering of buildings in Lower Manhattan show the years in which they were built from 1700 to the present. The gradations of dark orange to light orange correspond to the newest to the oldest buildings. The data used to create this map came from New York City's Department of City Planning.

MAPPING THE PHYSICAL ENVIRONMENT

DESIGN JL Cartography / Jonathan E. Levy   COUNTRY/REGION The USA

City Layouts

Topography, architecture and traffic routes give every city a unique structure. These conditions create the typical and individual inner structure of a city.

Luis Dilger didn't only want to show these structures in the conventional way from above. He also wanted to include the exact three-dimensionality of topography and buildings—a real world visualization. By importing satellite-based data into a 3D program, Dilger visualizes the fine lines and contours of architecture and streets. He then edits every detail individually until a minimalistic, yet realistic view of the city is achieved. These unique maps show some of the major cities of the world from an entirely new perspective. They represent a perfect mix of information and visual fascination.

DESIGN   Luis Dilger   COUNTRY/REGION   Germany

BOSTON

42° 21' 29" N, 71° 3' 49" W   population: 655,884   land area: 48.42 sq mi

MANHATTAN

40° 46' 0" N, 73° 59' 0" W
population: 1.634.795
land area: 59,5 km²

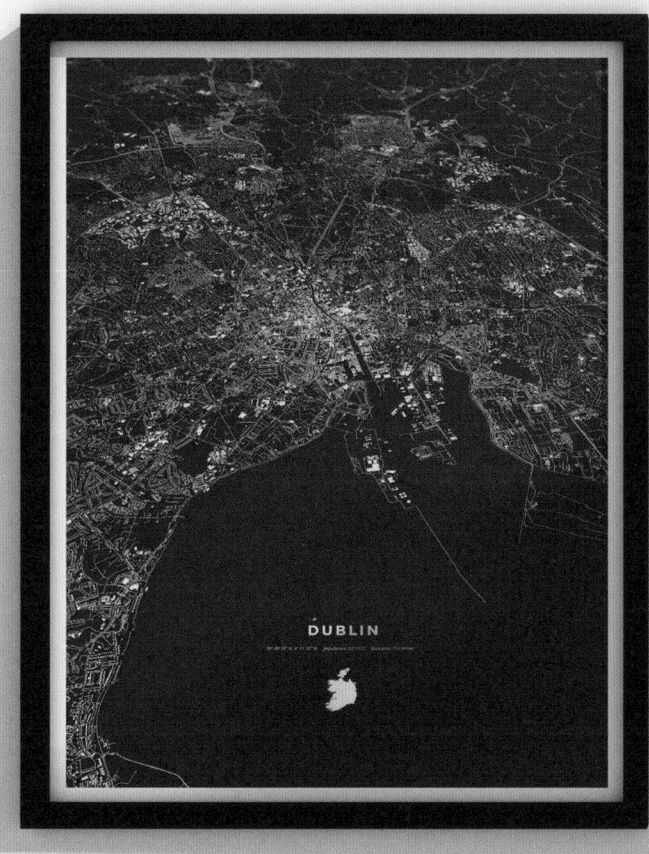

Yuansu Series

The Yuansu Series is a partnership between Chinese artist Ren Ri and bees. The art piece takes a form of relief sculpture.

Ren made an experiment when creating these works. He sought the help of bees, clearing one corner of the beehive and replacing it with relief maps. Amazingly, the bees accepted the invading object and continued shaping on the undulating terrain relief. Yet, they didn't cater to the choices human beings imposed on them, not entirely relying on the relief. They bit the high points of the relief and entered the space, continuing their honeycomb building whilst filling the lower ones with honeys based on their needs. The resulting maps are great and natural despite the "damage"—numerous holes—bees did on the maps. Actually, it's these holes that make the natural art pieces. The end result was a true collaboration between the artists and the bees.

PART 2

MAPPING HUMAN ACTIVITY

This section showcases maps portraying a variety of activities and covering topics including culture, economics, and politics.

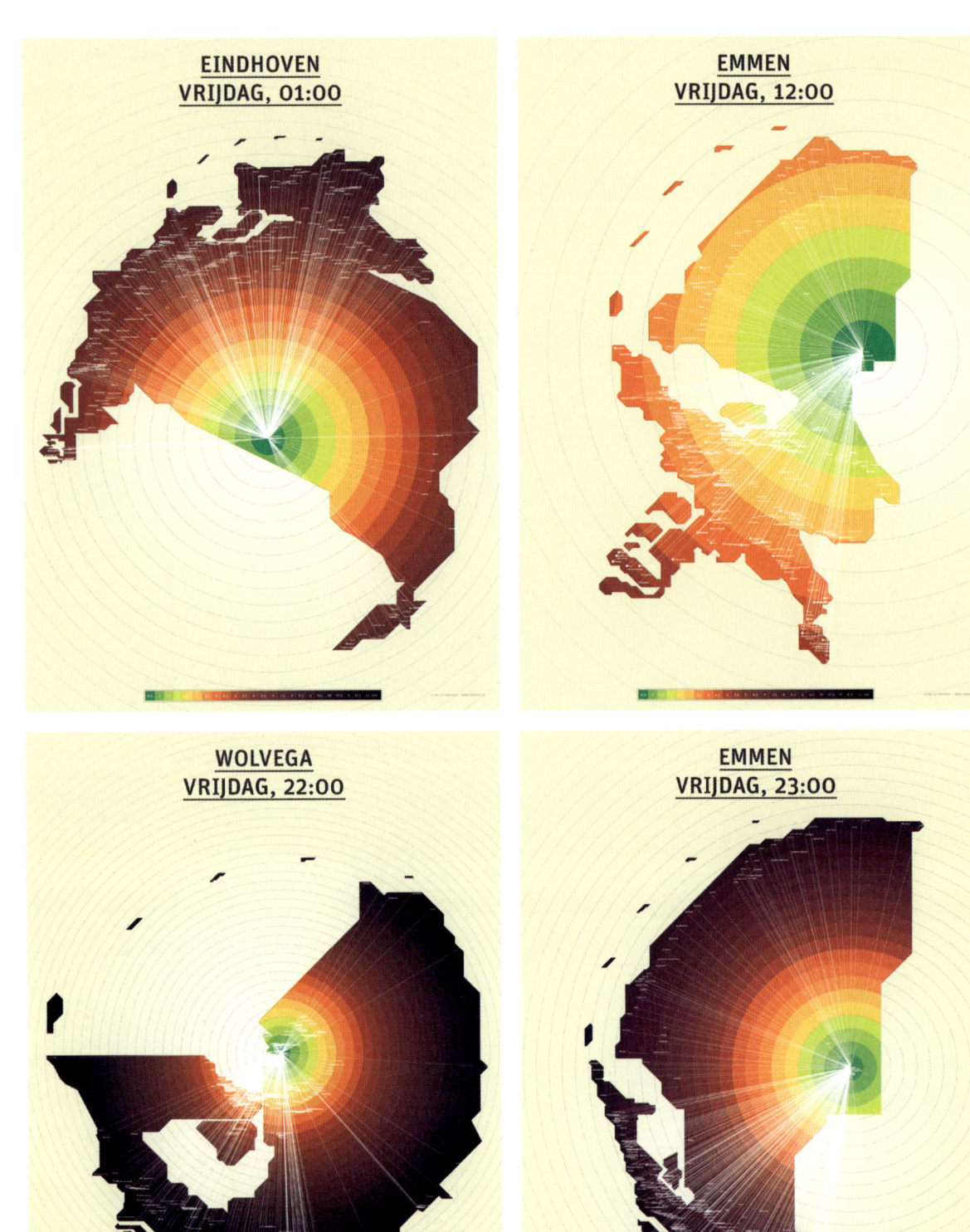

TimeMaps

Unlike traditional maps of the Netherlands, this interactive map focuses on time rather than actual distances among different locations within the country. In the Netherlands, trains bring people to their destinations easily and quickly; to some extent they decide the distance between two places. From the perspective of Amsterdam, for instance, the Netherlands is relatively small because of the quick and easy connections to other cities. But seen from a remote and small village such as Wolvega, the Netherlands is much bigger.
The hour of day is just as important as the location from which one travels. At night the map of the Netherlands would expand, because there are no trains, while in the morning it would shrink once trains commence their schedules. The map of the Netherlands will never be the same.

Tracking the Economic Disaster

Daniel Mason made this hardwood map to visualize the data of the economic crisis of 2008, trying to show the audience the overall impact of the crisis in a more straightforward way. He used the medium of wood and the method of elevation for this art piece. The elevation of the pieces is used to show the unemployment rate in a given county or state, and the color of the stain indicates the population density of the area. This piece was created for an audience in California, so it was designed to show the California context in the economic crisis at both a national and county level.

KING THE ECONOMIC DISASTER

DESIGN Daniel Mason   COUNTRY/REGION The USA

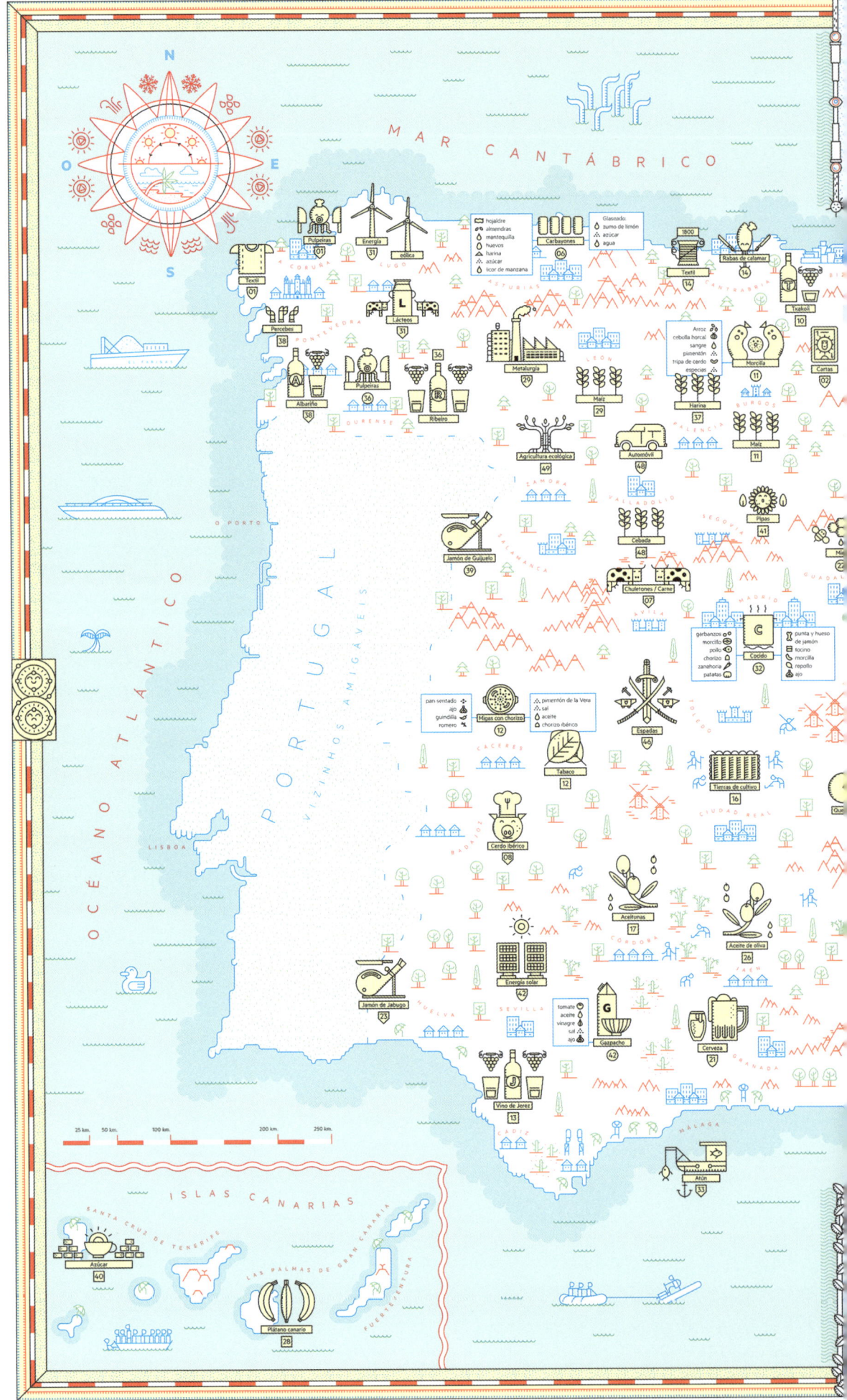

Goods of Spain Map

This full color illustrated poster was realized in collaboration with Walk With Me, a smart company that realizes and sells huge products! The idea was to represent the most important parts of Spanish production, from cocaine to ensaimada. A realistic representation of the country was based on the statistics and data collected by Walk With Me. The poster was enriched with further information such as a beautiful illustrated compass card, a seasons footer, and the constellations above Spain.

DESIGN relajaelcoco studio/ Pablo Galeano & Francesco Furno   COUNTRY/REGION Spain

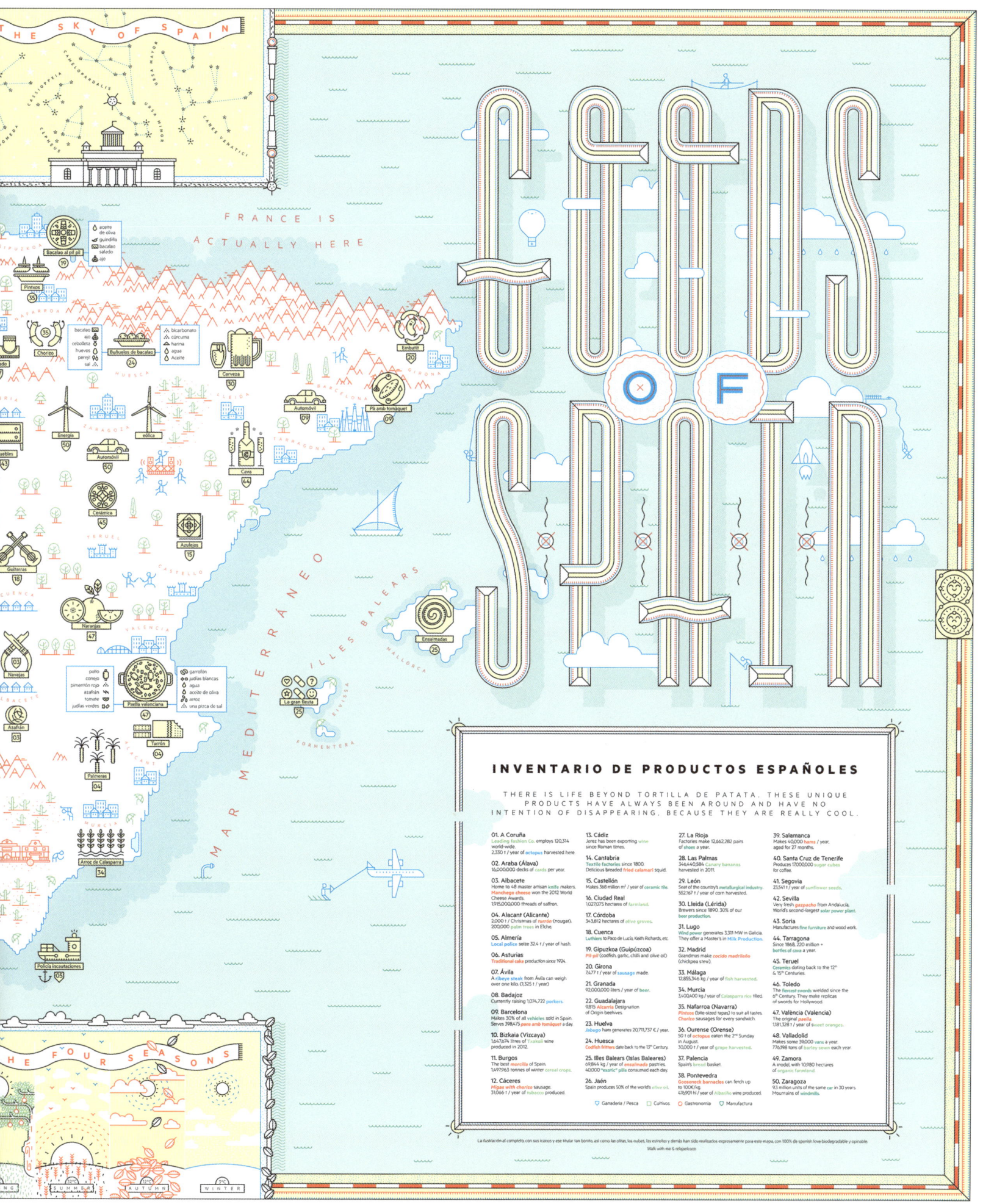

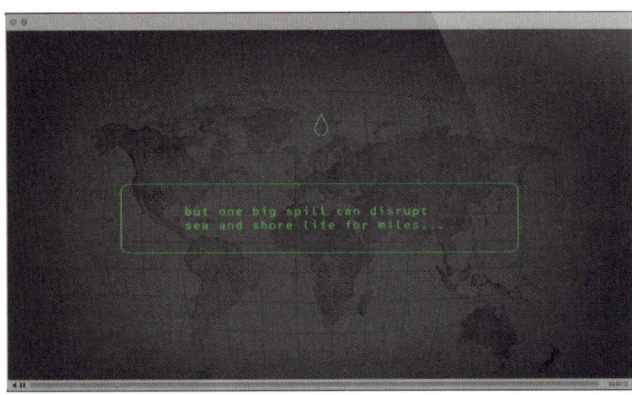
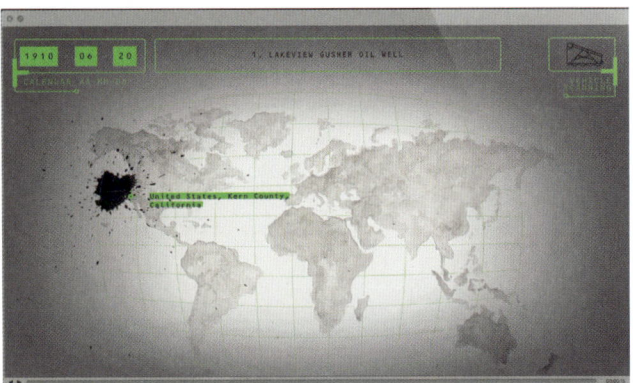

The Blackened Watercolor Map

The aim of this map was to visualize the pollution of oil spills. The designer represented the spilled oil as an ink drop dirtying the surface of a watercolor map. Colors give focus to the black drops. The designer furthermore presented a pollution spread process as a control panel of a Marine Radar Display. The dataset recorded the largest accidents between 1910 and 2010.

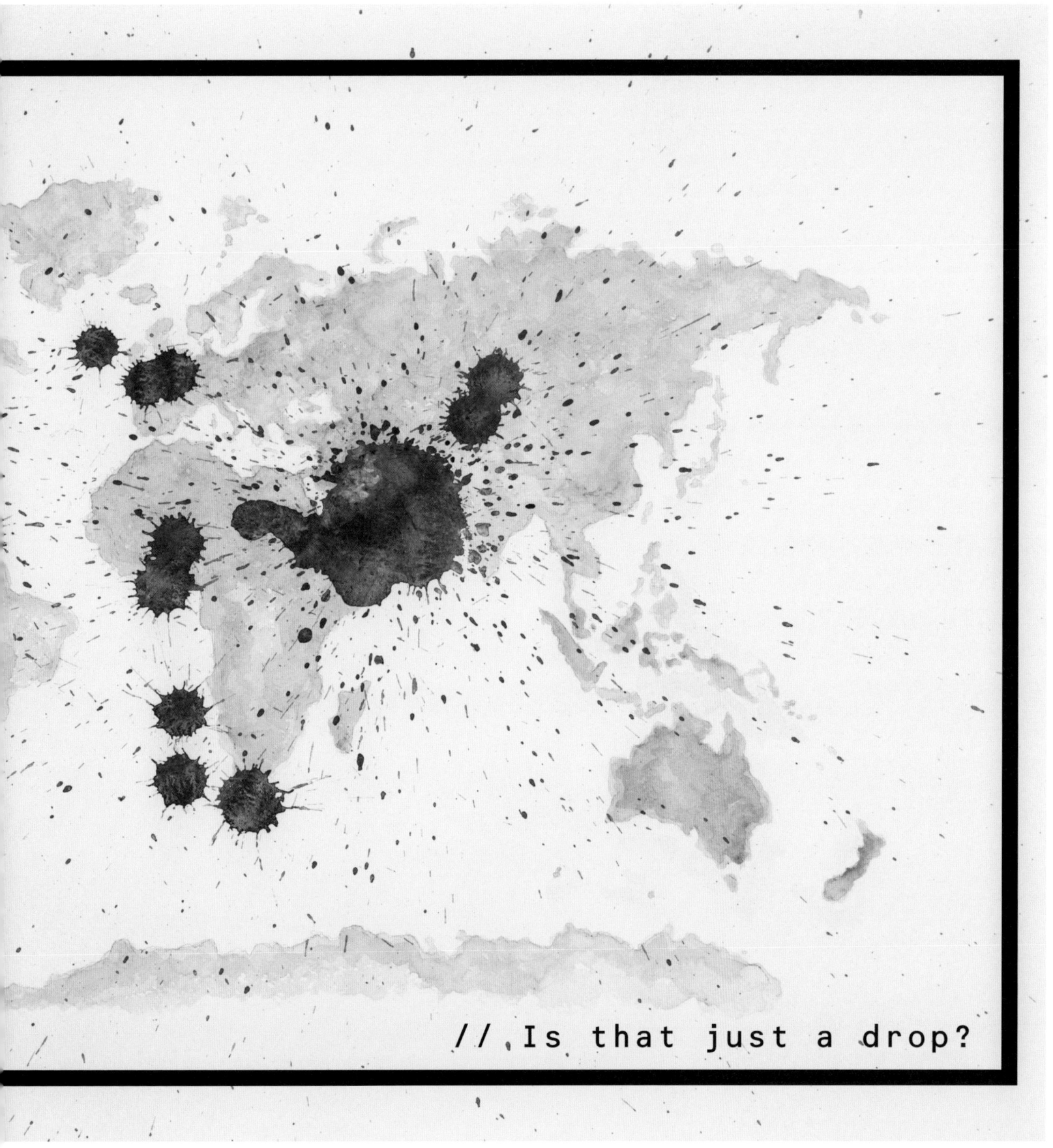

MAPPING HUMAN ACTIVITY

GP History Book

The map depicts the history of GP Group and its business activities around the world since its establishment. Its tone is colorful. The predominant green color of the illustration is consistent with the Group's logo color, enhancing a playful and light-hearted picture.

DESIGN Mandala Studio (Bangkok)   ILLUSTRATOR   Chinapat Yeukprasert   COUNTRY/REGION   Thailand

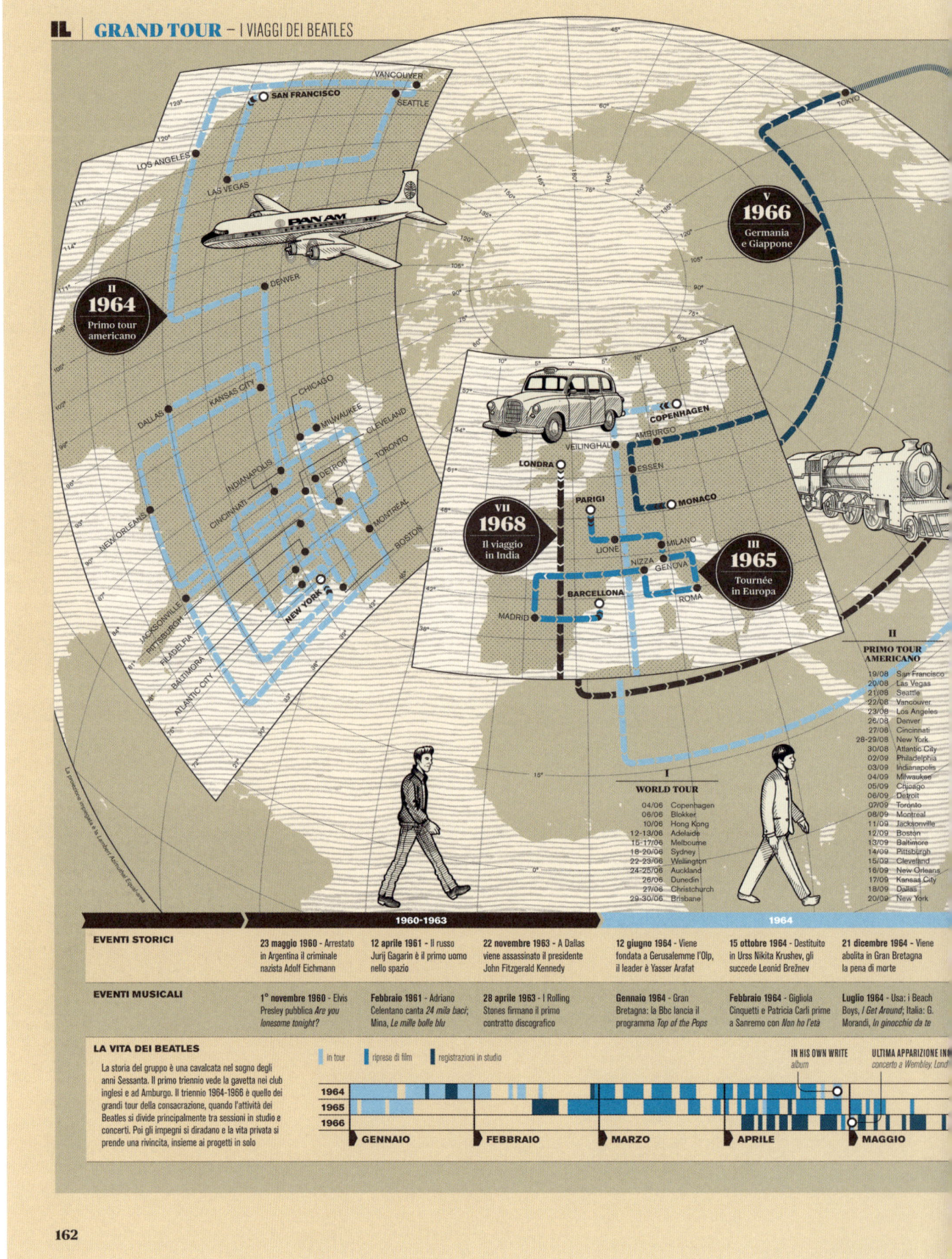

Grand Tour

Grand Tour is a section developed for IL, a monthly newsmagazine of the Italian financial newspaper Il Sole 24 ORE, from December 2009 to September 2011. It is a series of articles about the life of historical characters shown by infographic language, including Friedrich Nietzsche, Alexander von Humboldt, The Beatles, Primo Carnera and Giovanni Paolo II.

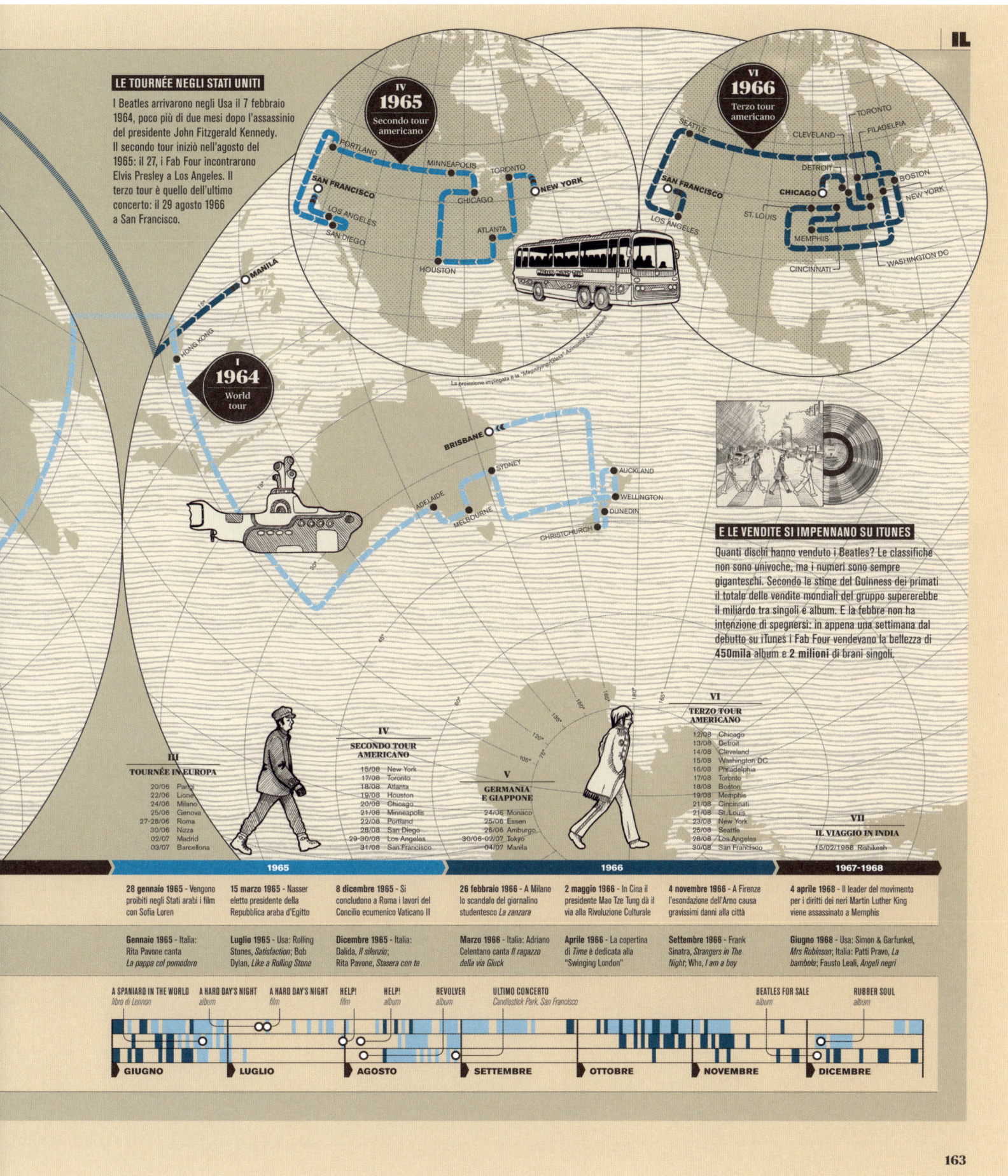

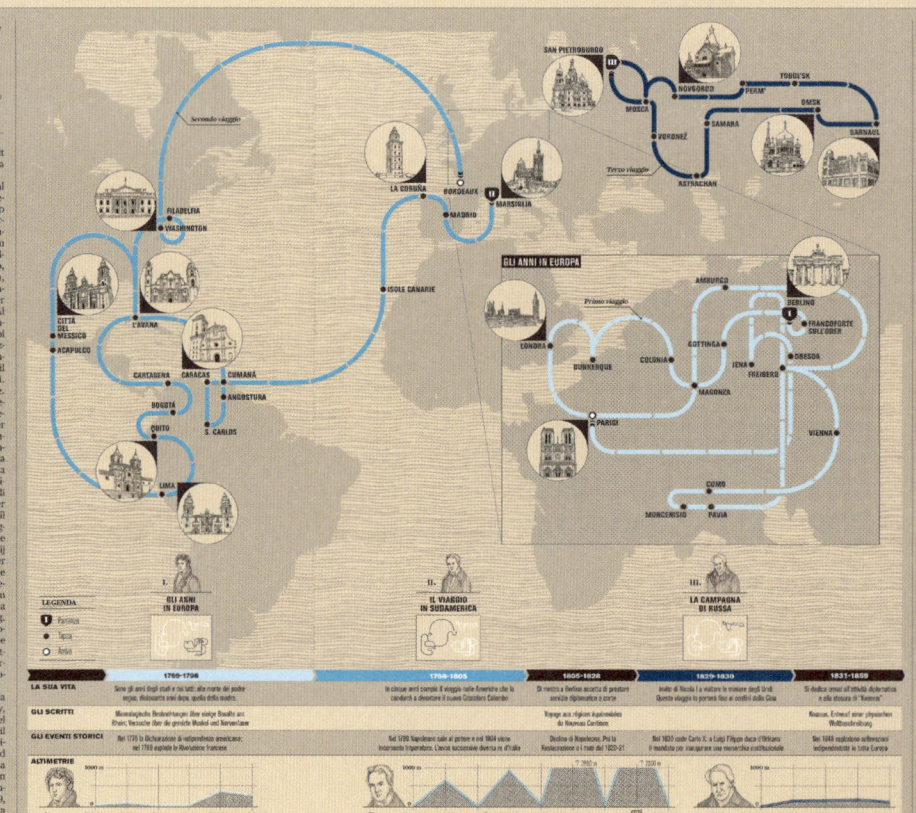

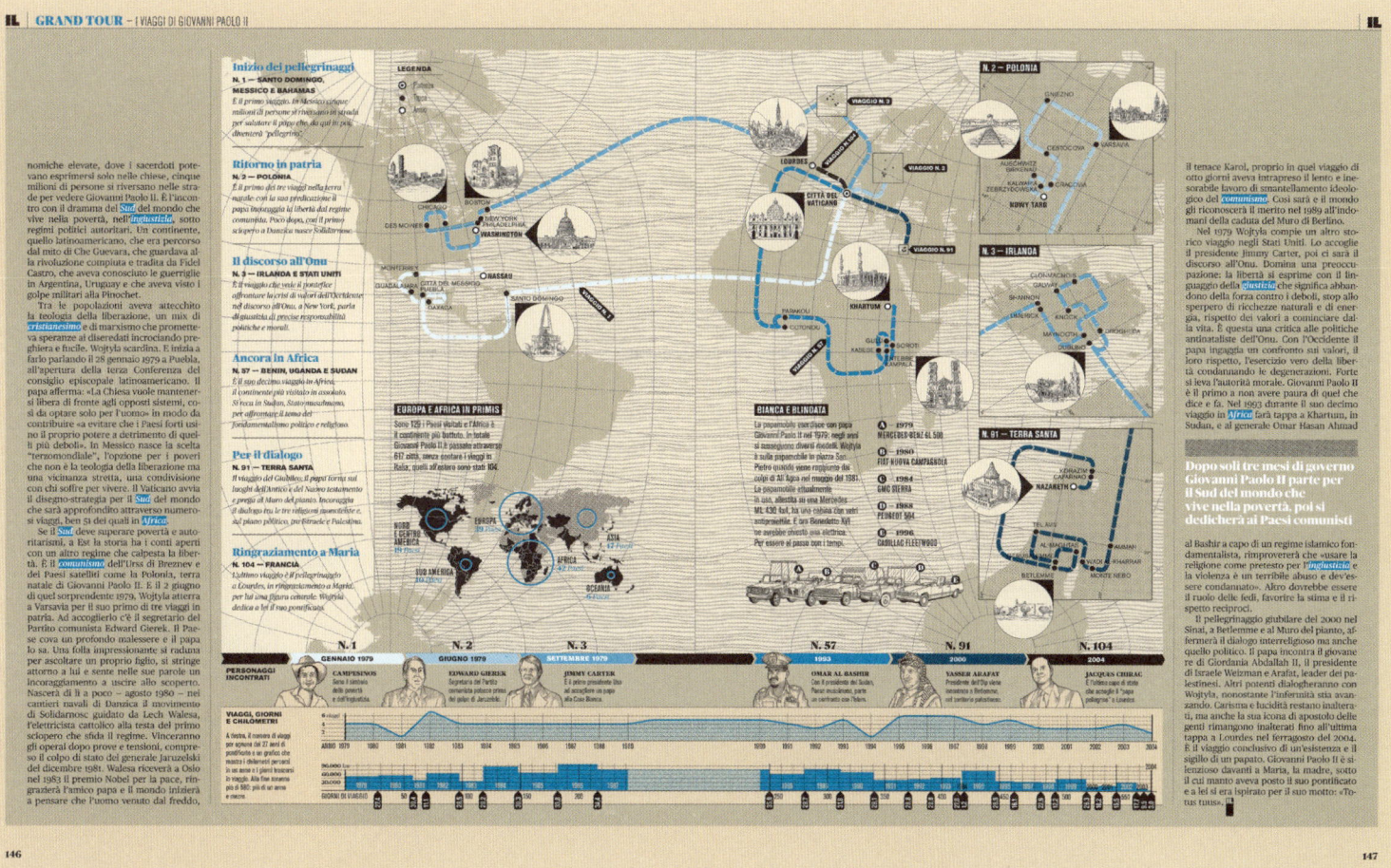

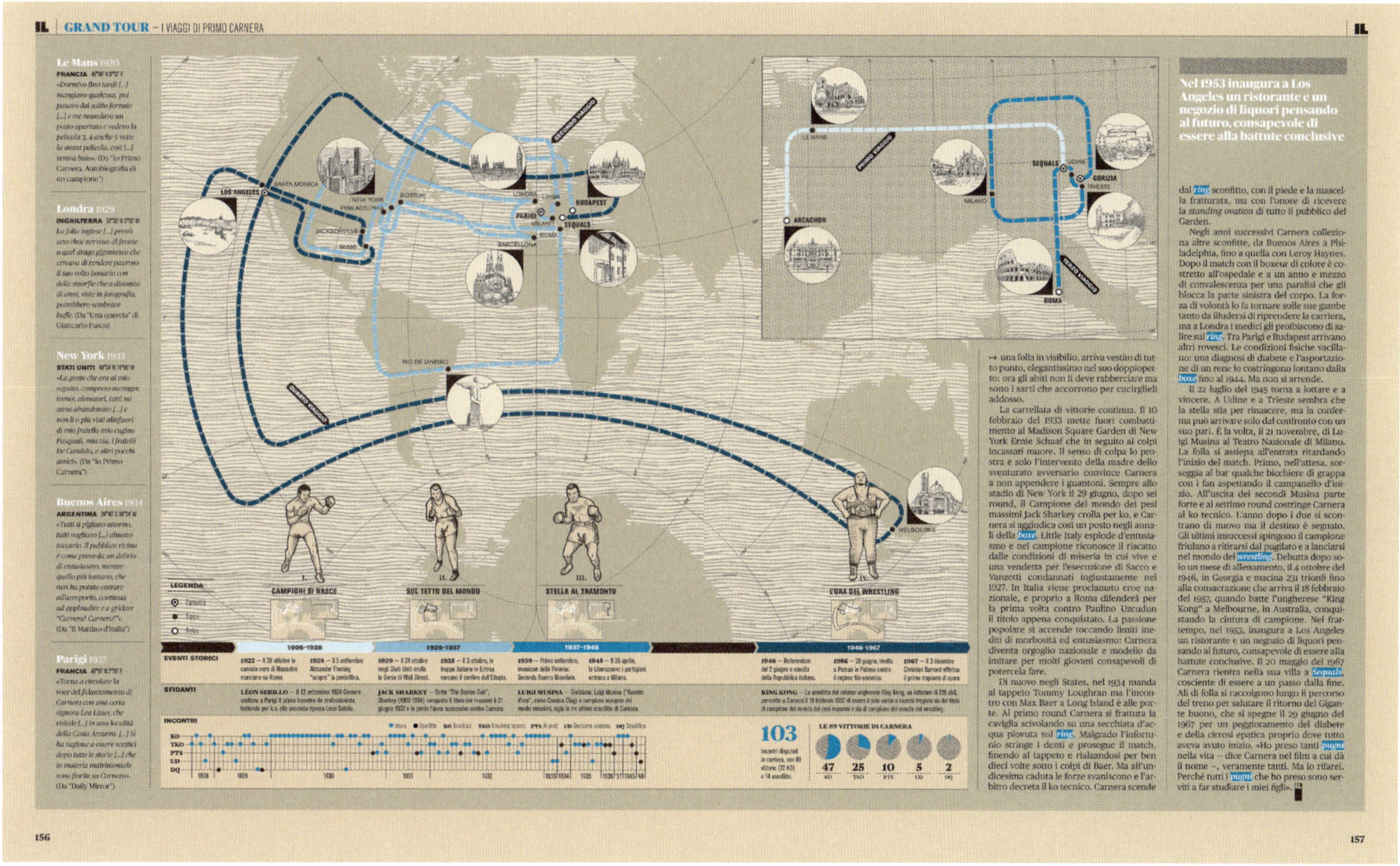

JUST RIDE
BIKE OFTEN

 Documentation of road and mountainbike rides in the DFW metroplex.
Information was collected through a span of three months.
All data was collected using the 'MountainBike' iPhone app.
Individual rides are depicted by a mountainform whose size corresponds to length of the ride.

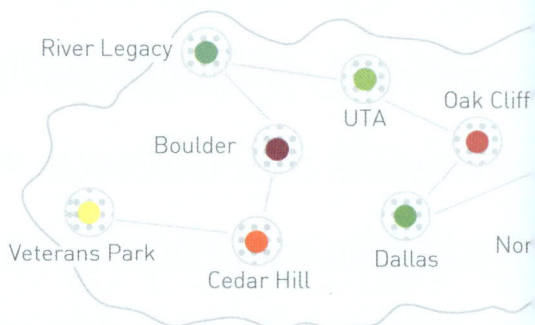

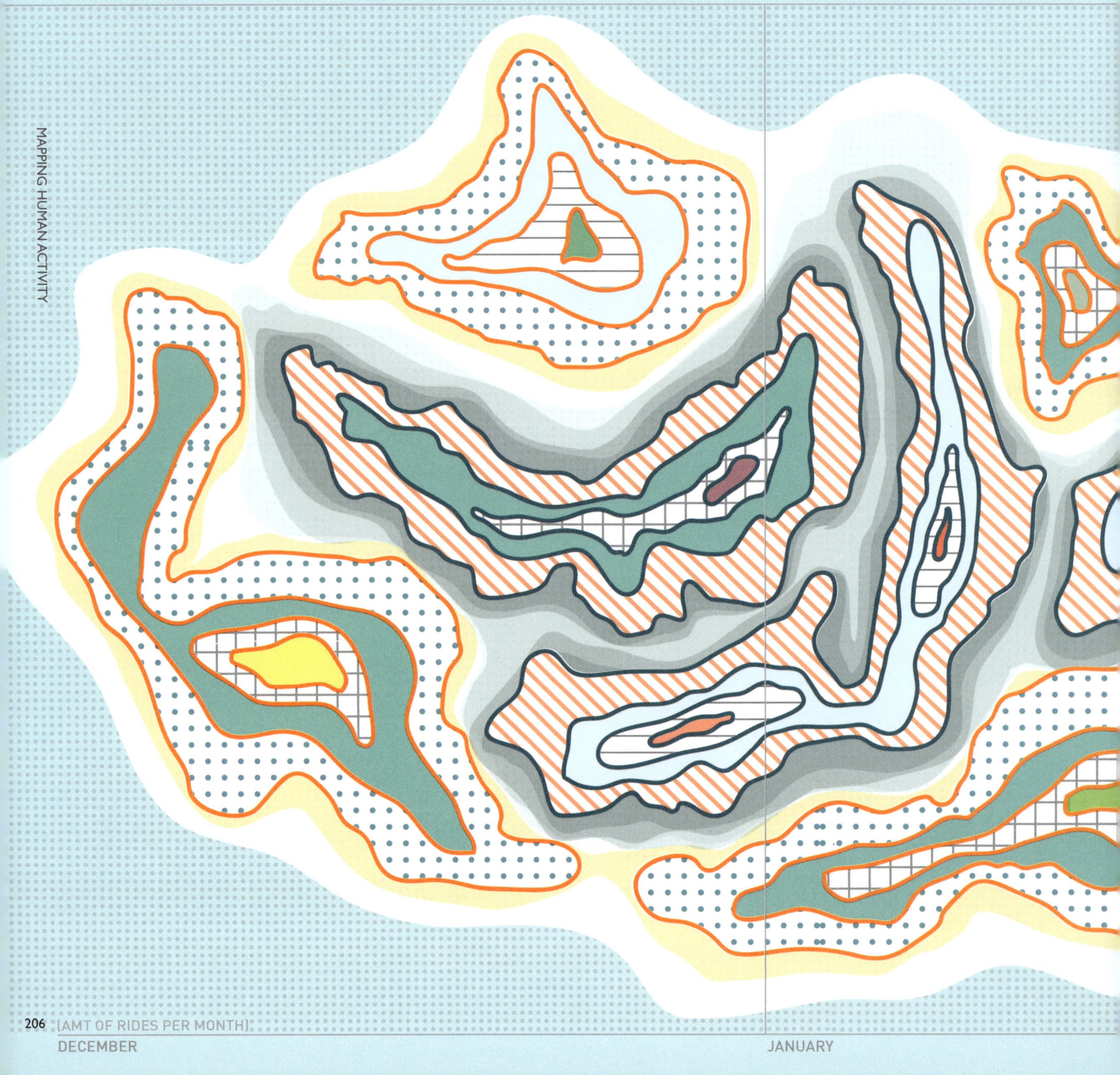

(AMT OF RIDES PER MONTH)
DECEMBER   JANUARY

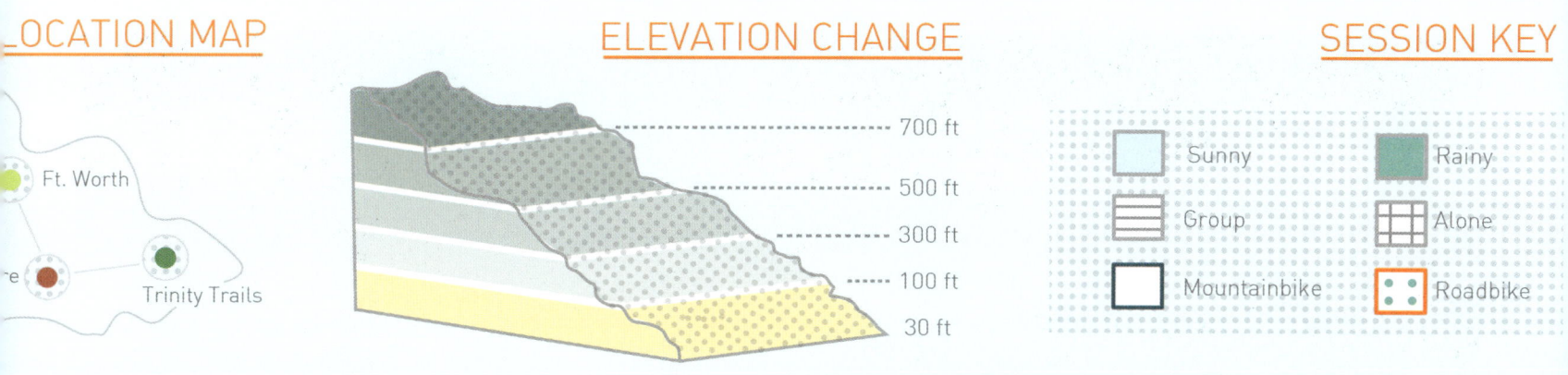

LOCATION MAP

Ft. Worth
Trinity Trails

ELEVATION CHANGE

700 ft
500 ft
300 ft
100 ft
30 ft

SESSION KEY

- Sunny
- Rainy
- Group
- Alone
- Mountainbike
- Roadbike

Data Visualization Map

This is a data visualization map which uses information generated from bike ride habits for a span of three months. The map shows multiple aspects and characteristics belonging to each individual ride. These characteristics include elevation change, overall location, weather, type of ride, and if the ride was completed in a group setting or alone. Individual rides are depicted by a mountain-form whose size also corresponds to the length of the ride. The timeline of months at the bottom of the map signifies a rough amount of ride-activity for each month, based on volume of mountains within the area. All data was collected using the MountainBike iPhone App.

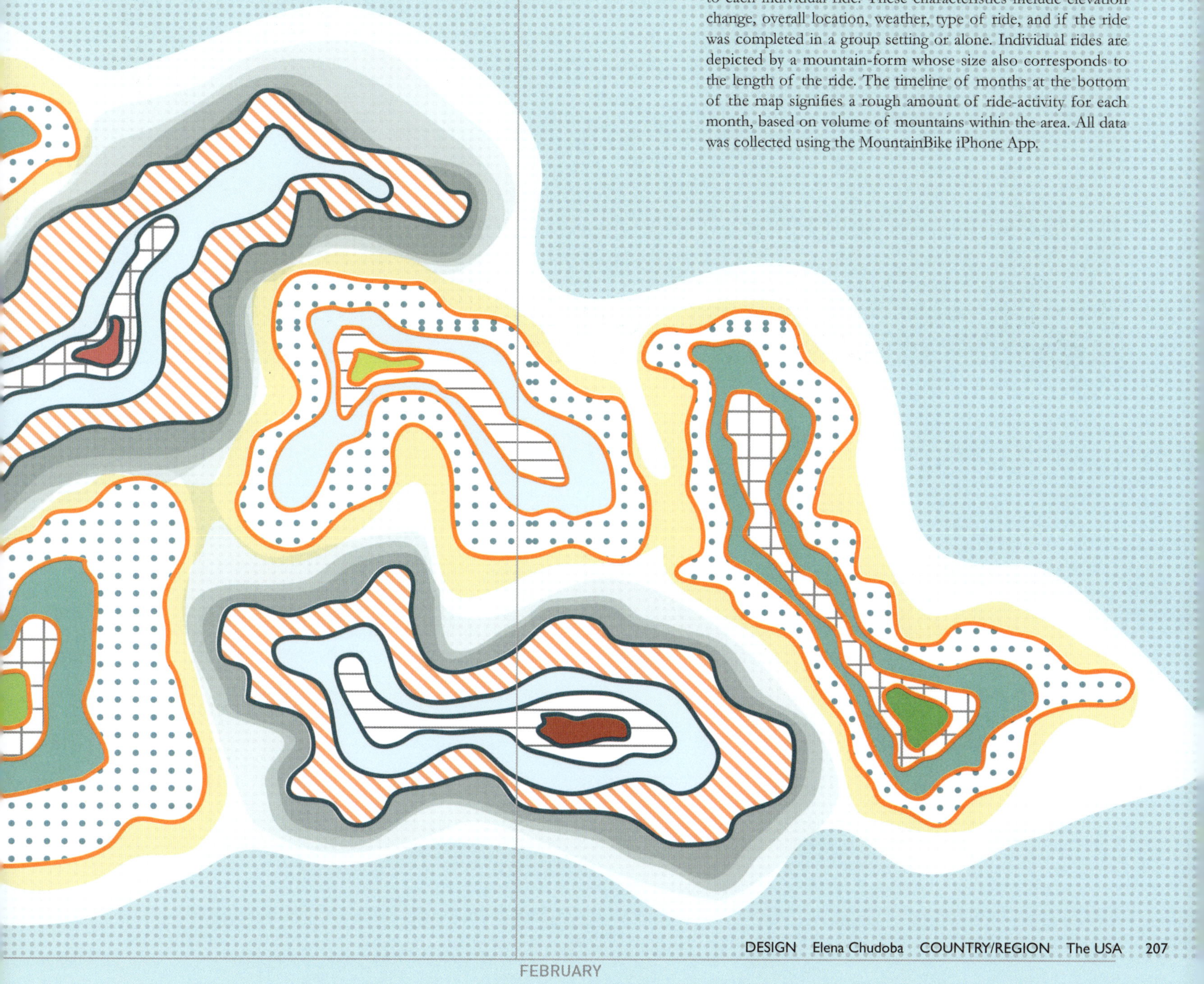

FEBRUARY

DESIGN Elena Chudoba COUNTRY/REGION The USA

Made in America STORY

STORY is a concept retail space that re-imagines itself every 4-to-8 weeks, shutting its doors to redesign and merchandise the entire space according to a new theme.

The theme for October 2013 was Made in America. On the inside wall, the "o" letter of STORY was transformed into an America map framed by luminous materials, and the store was divided into regions marked by a road map customers could follow along the walls. Each region was merchandised with products made in the corresponding region of the United States. The Midwest section of the store carried products made in the Midwest, etc. This project applies map design to business activity.

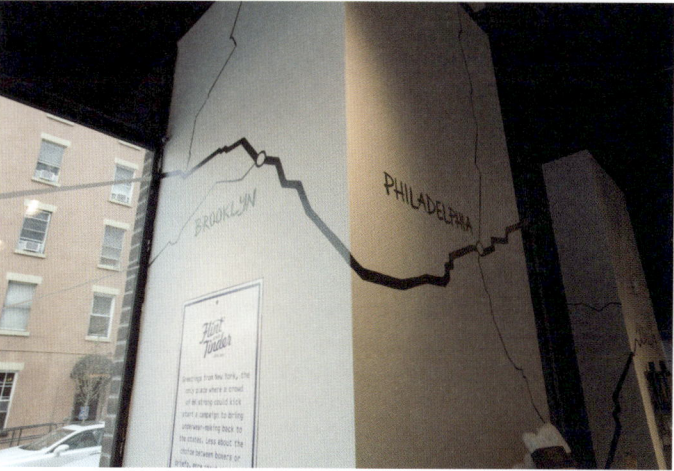

Ecumenopolis —Mapping Project

This collage is a mapping of Savic's friend M.F.'s voyages during the past decade. The materials used came from his collected items during the travels—city maps, subway maps, transport cards, tickets, etc. It may be seen as a portrait of M.F., but also as a map of an imaginary city—Ecumenopolis, which was created from the fragments of different cities from all over the world.

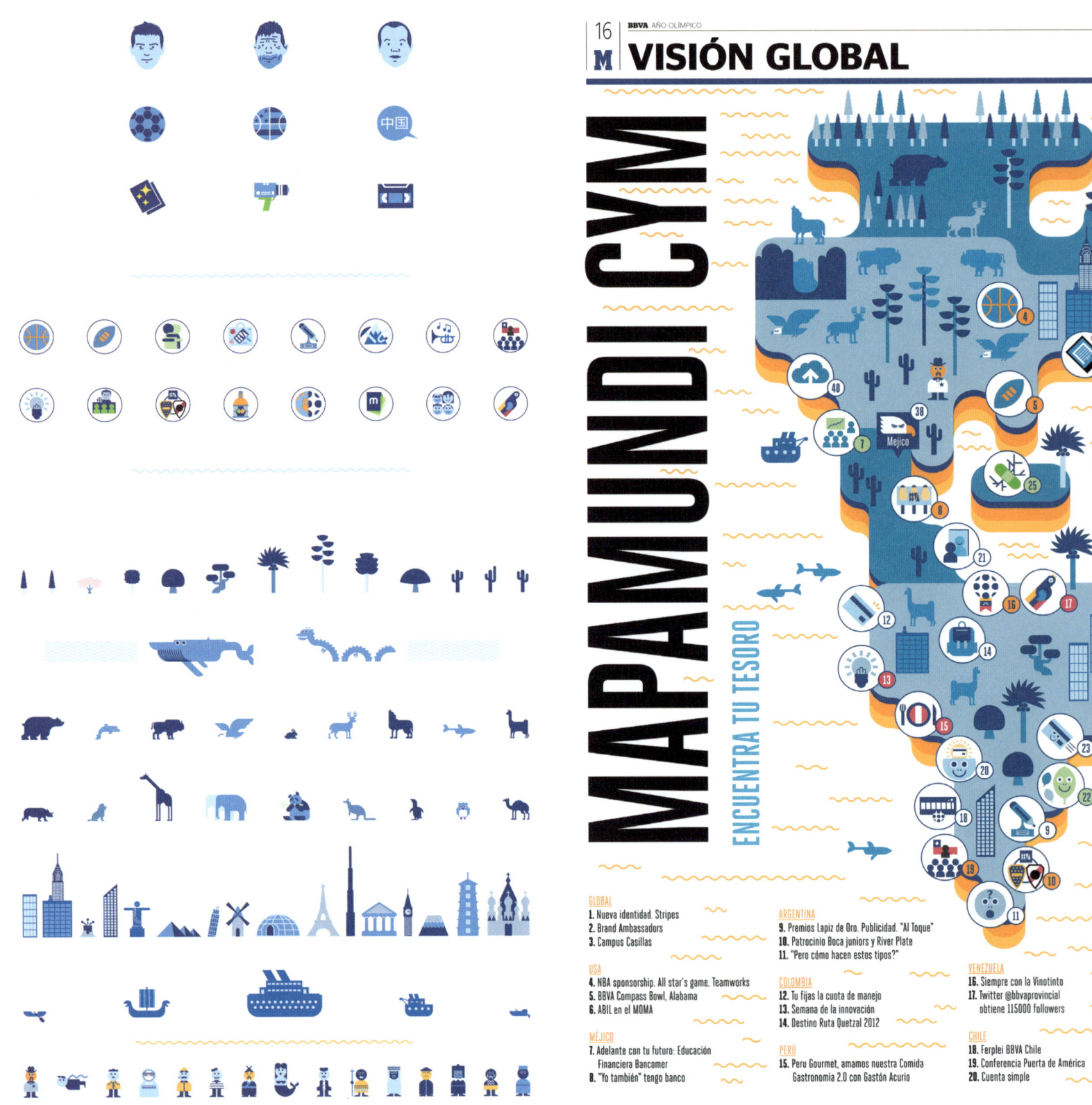

BBVA Infographics

This was a strong editorial project made for Marca BBVA, an internal issue about BBVA life. Two double-page infographics based on BBVA data were used to illustrate the best hits of 2012.
The second double-page is a map of the world with the most important goals reached in 2012 by the BBVA Company worldwide. Relajaelcoco studio developed everything using the corporate color palette of the bank—blue, but in different shades. And the vector system is inspired by the BBVA logo shapes. A strong visual result is the outcome.

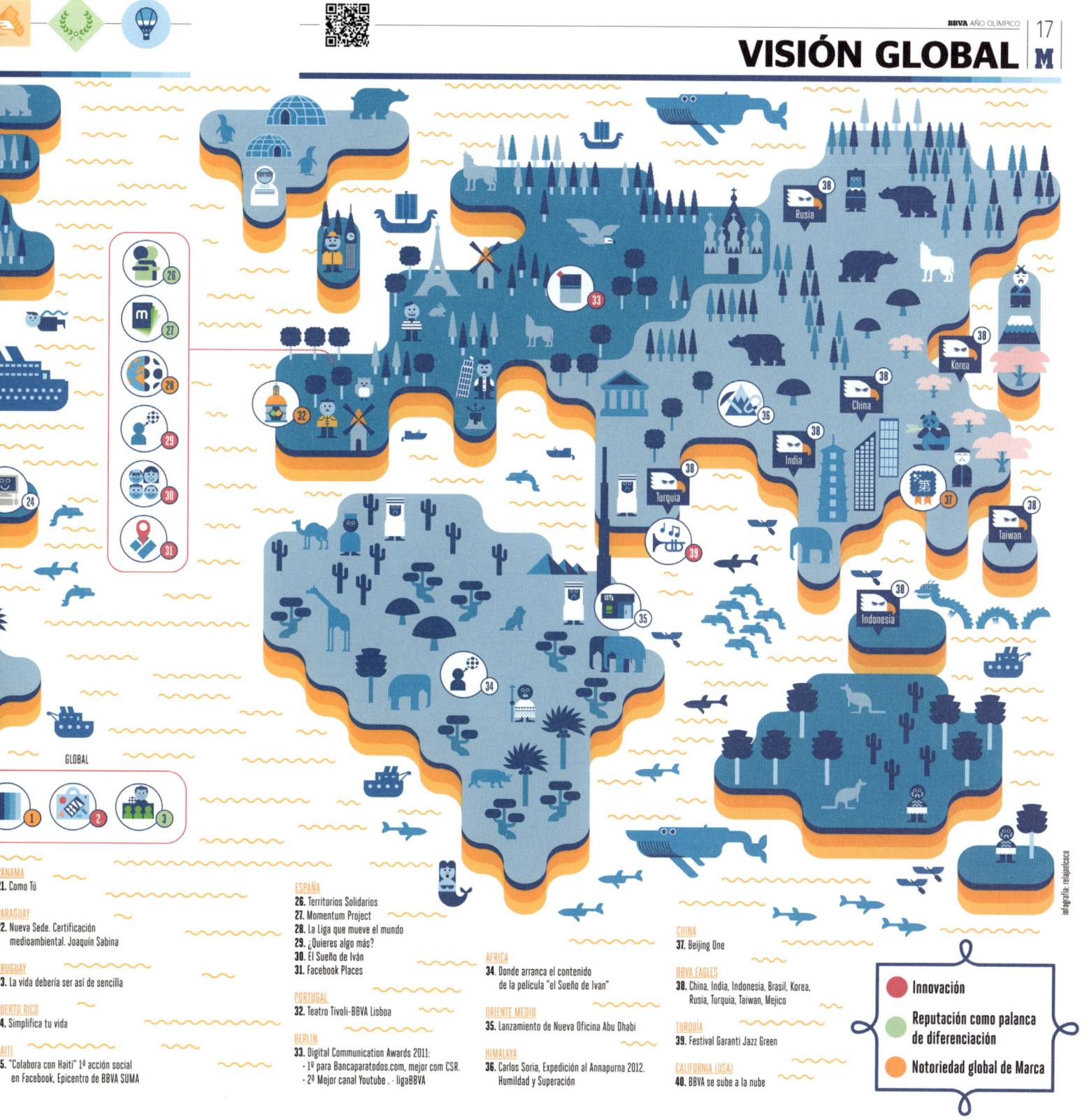

A SUBJECTIVE MAP OF NEW YORK CITY

A Subjective Map of New York City

This is a personal map of New York City showing all of Meertens and his partner's movements during one year. They tracked their location with OpenPaths, blue representing the designer himself and red for his partner. The yellow dots are locations where they took photos. When combined, a subjective map emerges from their movements in the city. All 10,760 datapoints were collected from March 2012 to January 2013.

DESIGN Vincent Meertens   COUNTRY/REGION The Netherlands

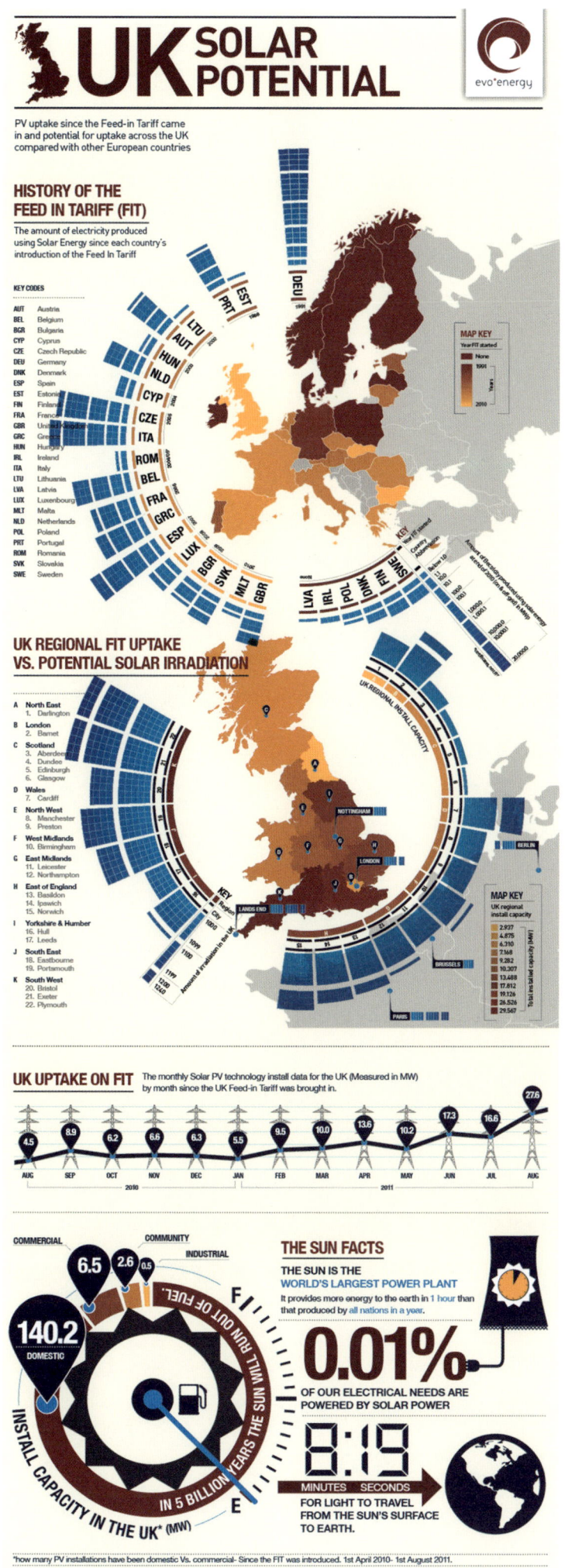

UK Solar Potential

The data visualization shows the UK's solar potential, comparing regions and countries that have taken up the government feed-in tariff from 1991 to 2010.

DESIGN Leigh Riley   COUNTRY/REGION The UK

MAPPING HUMAN ACTIVITY

101 New York Sights Map

Rod Hunt was commissioned to illustrate the 101 New York Sights map for Circle Line Sightseeing Cruises. The key aims were to show all of the 101 New York landmarks seen on the full island circumnavigation of Manhattan, while keeping the map uncluttered. A bold graphic simplified approach to depicting Manhattan Island with detailed buildings and attractions was chosen as the best solution.

216　DESIGN　Rod Hunt　COUNTRY/REGION　The UK

_042

Geografia Dell' Amianto

The map shows the correlation between asbestos contamination and the cancer mesothelioma. The density of red lines measures the mortality caused by mesothelioma. The blue dots denote the asbestos-polluted areas reported by the Italian government.

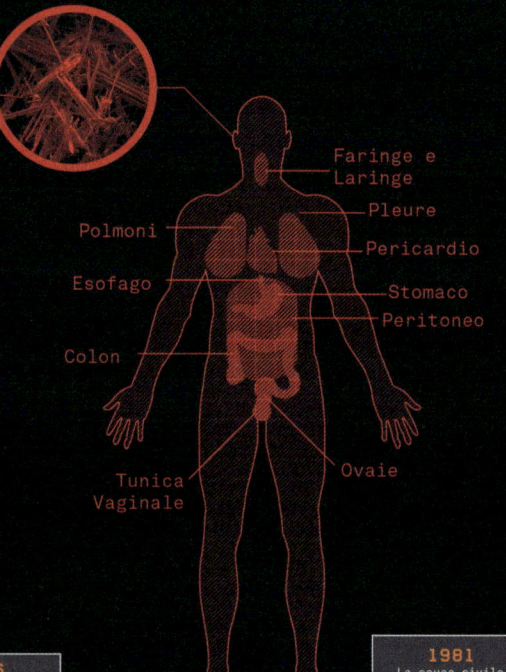

COME SI LEGGE

La mappa a sinistra mostra i 38mila siti contaminati censiti dal ministero dell'Ambiente (i punti blu) sullo sfondo della mortalità per provincia per mesotelioma pleurico (il tumore dell'amianto) e dei 12 Sin, i Siti di interesse nazionale dove la bonifica è più urgente. Anche se parziale (le nostre stime parlano di 300mila siti e non sono incluse le morti per malattie asbesto-correlate), questa è, a oggi, la miglior fotografia dell'esposizione all'amianto in Italia. Il dataset è scaricabile su Wired.it.

LEGENDA

Mortalità (per 100mila abitanti)
8 6 5 4 2

AREE A RISCHIO
- Segnalato dal ministero dell'Ambiente
- Sito di interesse nazionale (SIN)

Siti:

- Balangero — 3.100.000 m²
- Emarése — 400.000 m²
- Broni — 135.000 m²
- Casale Monferrato — 740.000.000 m²
- Pitelli — 18.830.000 m²
- Massa Carrara — 27.000.000 m²
- Area litorale vesuviano — 81.210.000 m²
- Tito — 60.000 m²
- Bari — 100.000 m²
- Aree industriali Val Basento — 340.000.000 m²
- Priolo — 33.500.000 m²
- Biancavilla — 3.300.000 m²

2014 — La Cassazione prescrive il reato di disastro ambientale per Eternit assolvendo il magnate svizzero Stephan Schmidheiny.

2015 — Il 12 maggio si apre la prima udienza del processo penale Eternit Bis con imputazioni per omicidio.

MORTI PER MESOTELIOMA, IL TUMORE PIÙ DIFFUSO CAUSATO DALL'AMIANTO

*non sono conteggiati i casi di Lombardia, Valle d'Aosta, Abruzzo, Calabria, Sardegna e Molise

DECESSI EFFETTIVI* | DECESSI EFFETTIVI IN TUTTA ITALIA | PROIEZIONE DECESSI FUTURI

1993 1994 1995 1996 1997 1998 1999 2000 2001 2002 2003 2004 2005 2006 2007 2008

FONTI: ISTAT, MINISTERO DELL'AMBIENTE
FONTI: RENAM, ISTAT · VISUALIZZAZIONE DATI MASSIMILIANO MAURO

National Scout Jamboree Map

Boy's Life, the official publication of The Boy Scouts of America, commissioned Rod Hunt to illustrate the map for the 2013 National Scout Jamboree that took place at the Scott Summit Center, USA. The aim was to create a fun and engaging map for use as a navigational tool while on site, showing all the activities available and locations of key facilities.

MAPPING HUMAN ACTIVITY

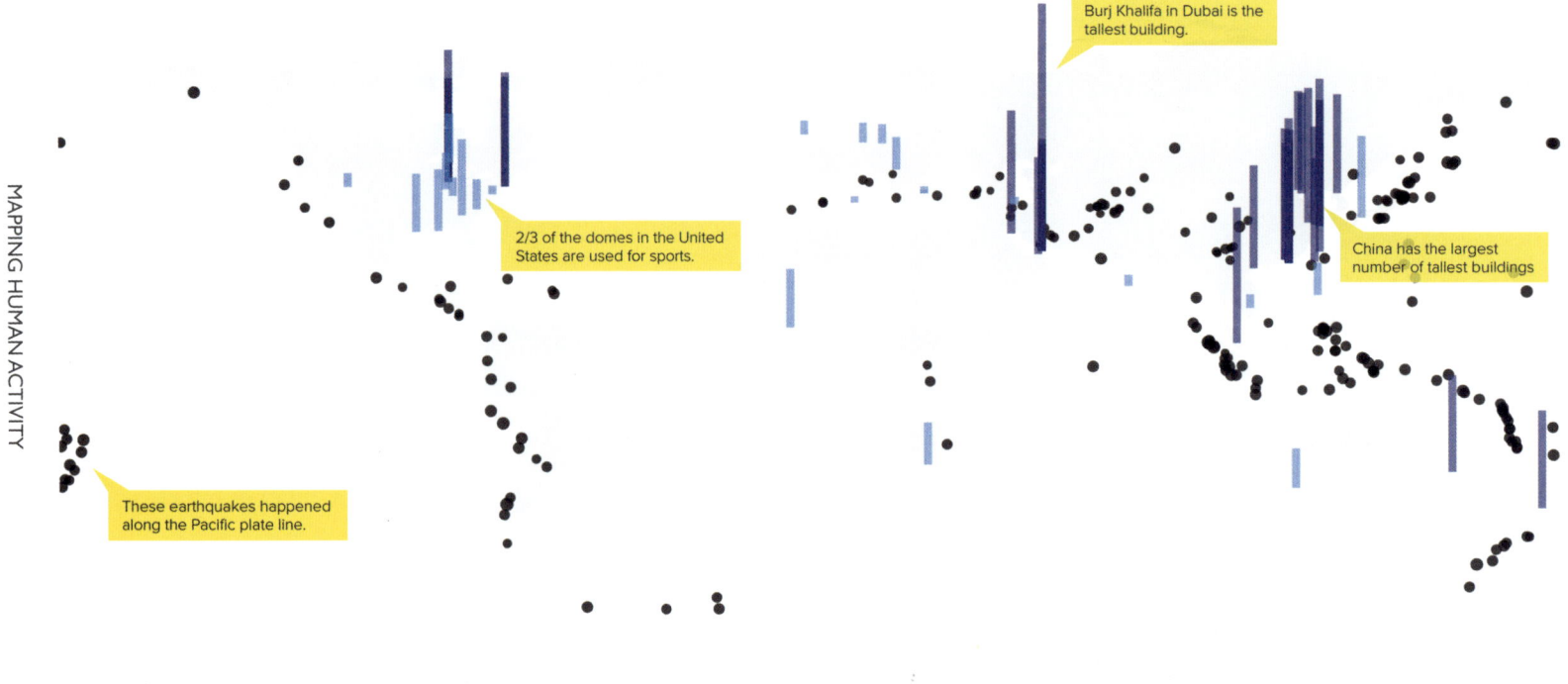

 1900 TO 2011

TOTAL
EARTHQUAKES
MAPPED BY NUMBER

TALLEST
BUILDINGS
MEASURED IN HEIGHT

LARGEST
DOMES
MEASURED IN DIAMETER

Earthquakes, Buildings and Domes

This map was created to find out how some of the most phenomenal physical structures would compare to the amount and location of earthquakes that have happened in the last century. Circles reflect the number of earthquakes, while bar charts reflect the height of buildings and diameter of domes.

DESIGN Carrie Winfrey   COUNTRY/REGION   The USA

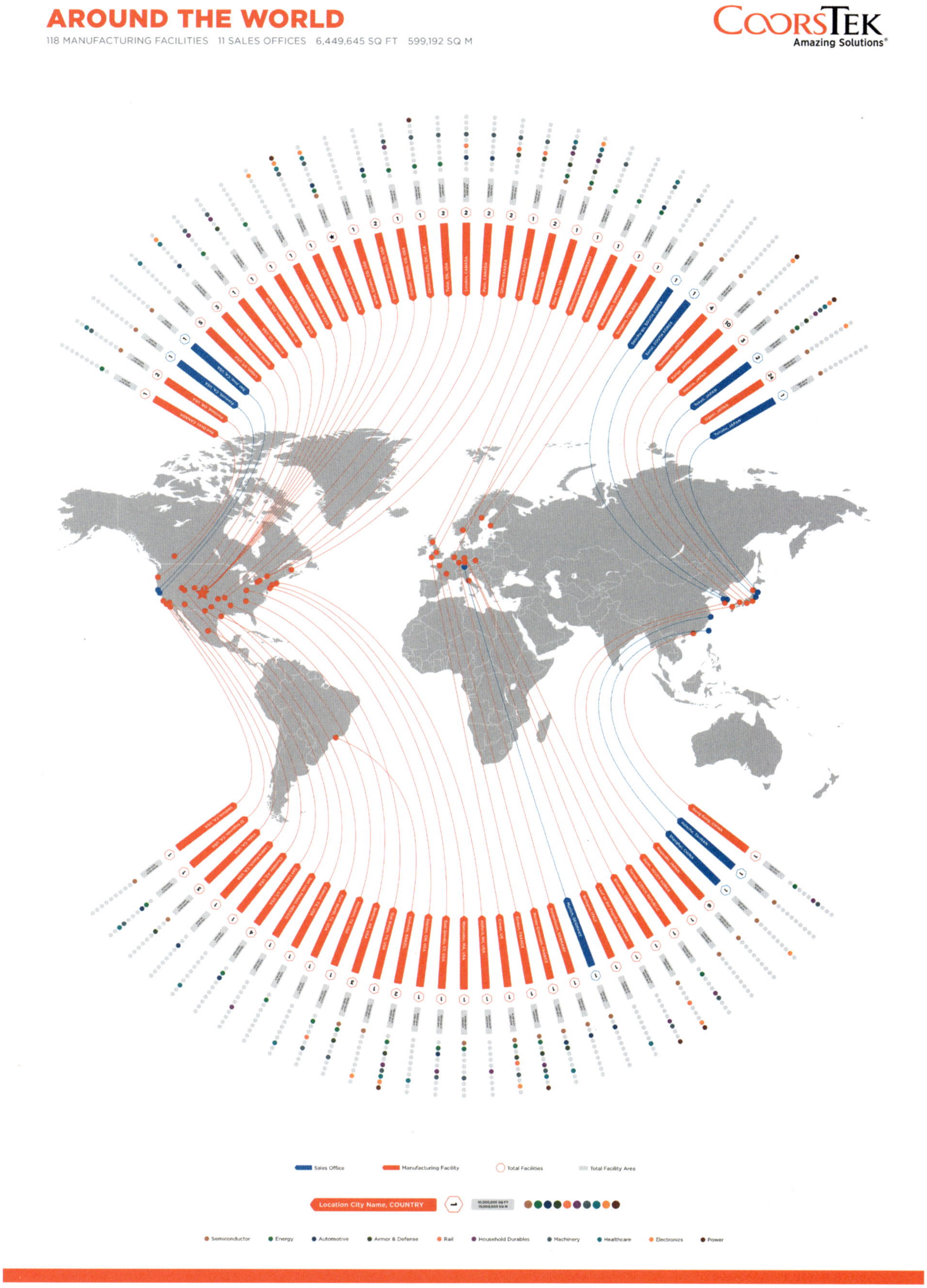

AROUND THE WORLD

118 MANUFACTURING FACILITIES   11 SALES OFFICES   6,449,645 SQ FT   599,192 SQ M

Around the World

The goal of this poster was to highlight the global presence of CoorsTek, making its employees and customers aware of its size and global reach. The finished poster was sent around the world to each facility to hang in their lobbies and other high-traffic areas, showing off the size of the company, its manufacturing capacity, and the variety of industries that it serves around the world.

DESIGN   CoorsTek, Inc. / Will Manning   COUNTRY/REGION   The USA

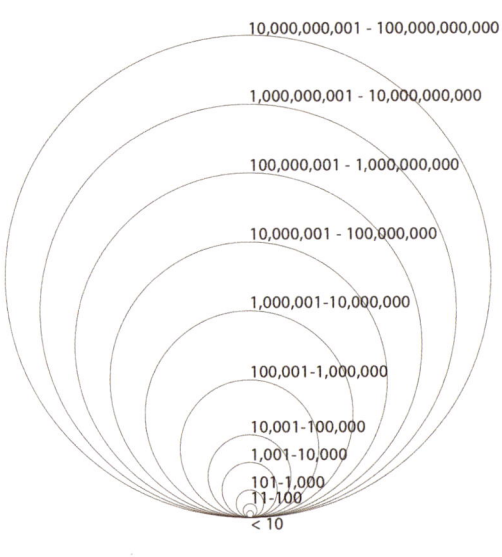

NUMBER OF PEOPLE RECORDED

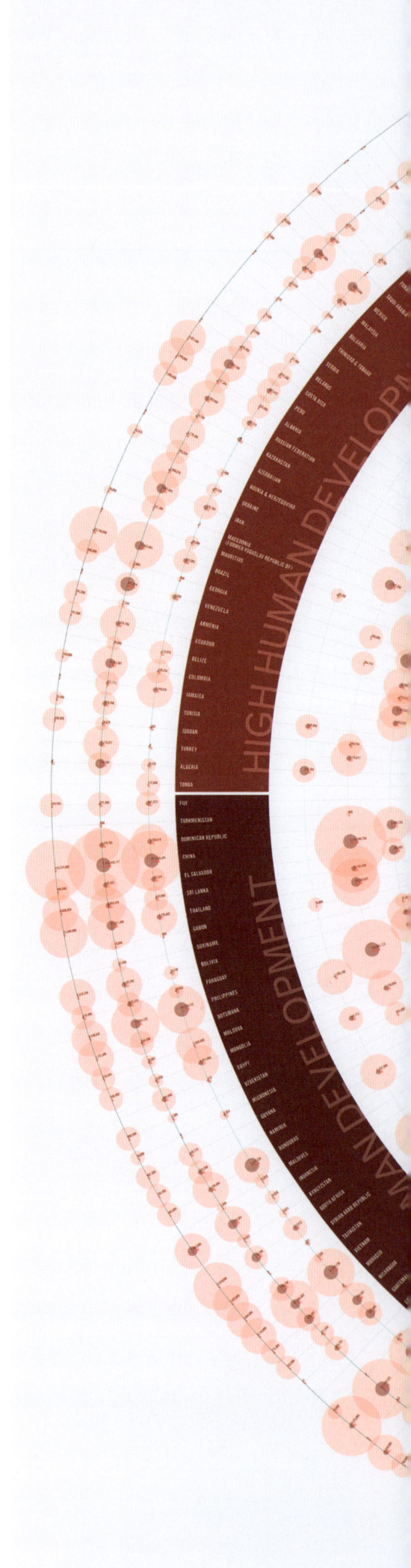

MAPPING HUMAN ACTIVITY

UNDP Project

This information graphic provides a visual reference for data previously only depicted in numeric spreadsheets provided by scientists and governments, plotting the alarming number of people affected and killed by water-related natural disasters for a time period of 30 years. It also reveals those countries most in need of disaster recovery resources.

224   DESIGN Rhiannon Fox   COUNTRY/REGION Bermuda

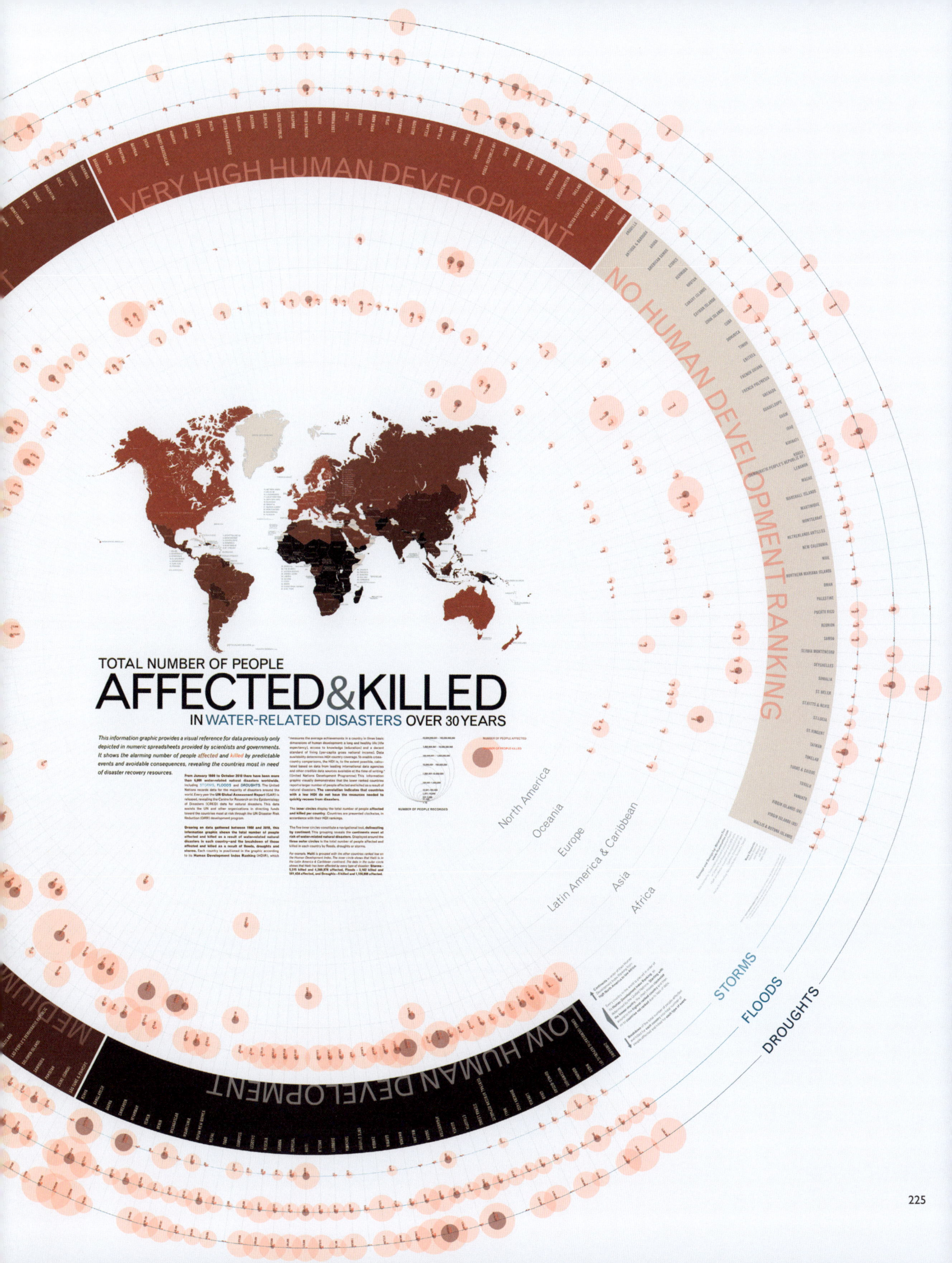

Atlas of Public Spaces in Urbino/Controlled Areas

Controlled Areas is a part of The Atlas of Public Spaces in Urbino, a project analyzing different aspects of public space in Urbino led by Joost Grootens at ISIA Urbino. This part focused on mapping the video surveillance areas of public spaces, in particular the old town center of Urbino. The video surveillances were divided on the strength of typology and category. Observing the map, viewers can see clearly where surveillance is crowded and the comparatively absent areas.

DESIGN Arianna Di Betta, Veronica Maccari   COUNTRY/REGION Italy

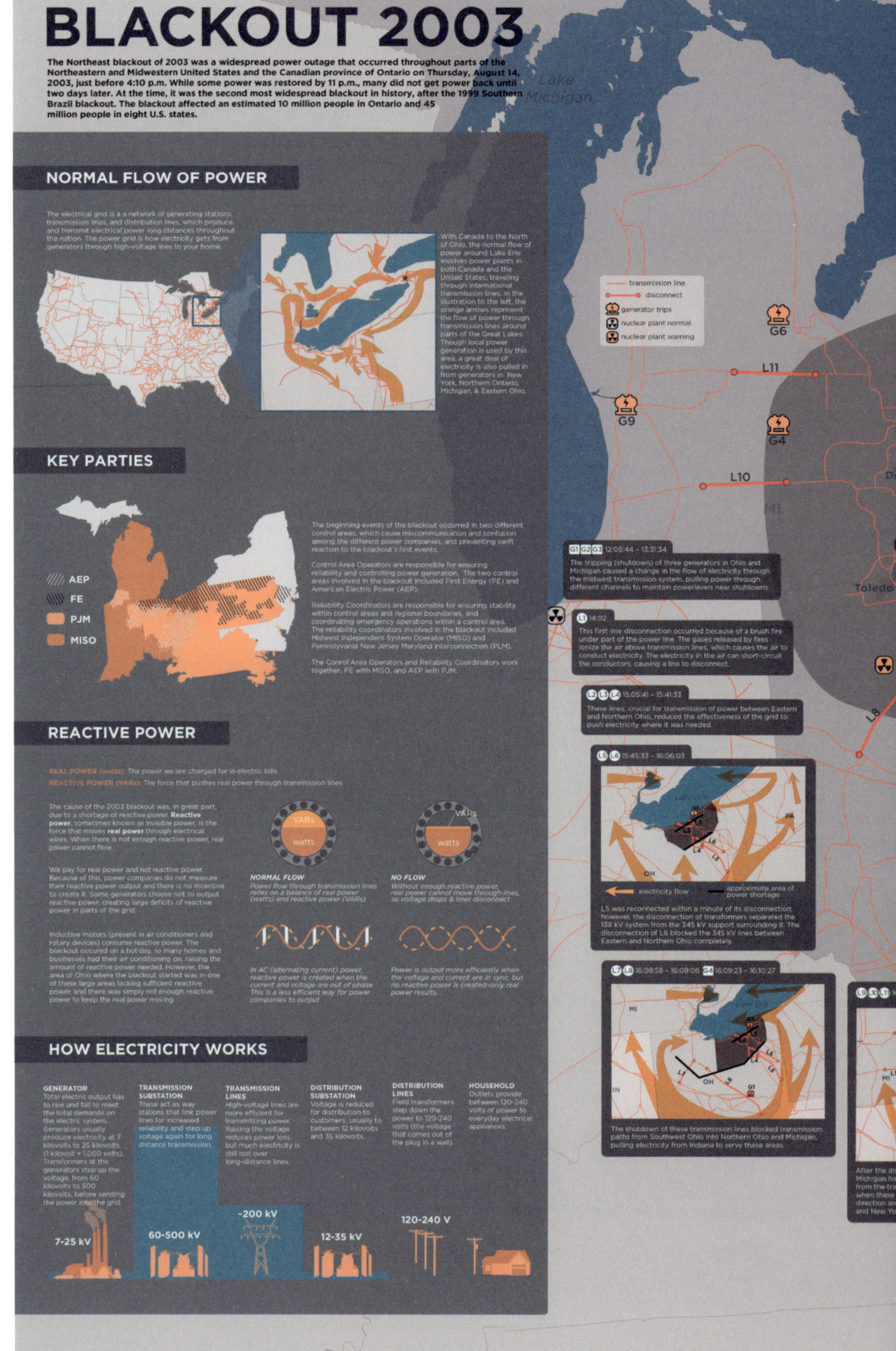

Blackout 2003

This project is a narrative infographic of the worst blackout in U.S. history. The Northeast blackout of 2003 was a widespread power outage that occurred throughout parts of the Northeastern and Midwestern United States and the Canadian province of Ontario on Thursday, August 14, 2003, just before 4:10 p.m. While some power was restored by 11 p.m., many did not get power back until two days later. At the time, it was the second most widespread blackout in history, after the 1999 Southern Brazil blackout. The blackout affected an estimated 10 million people in Ontario and 45 million people in eight U.S. states.

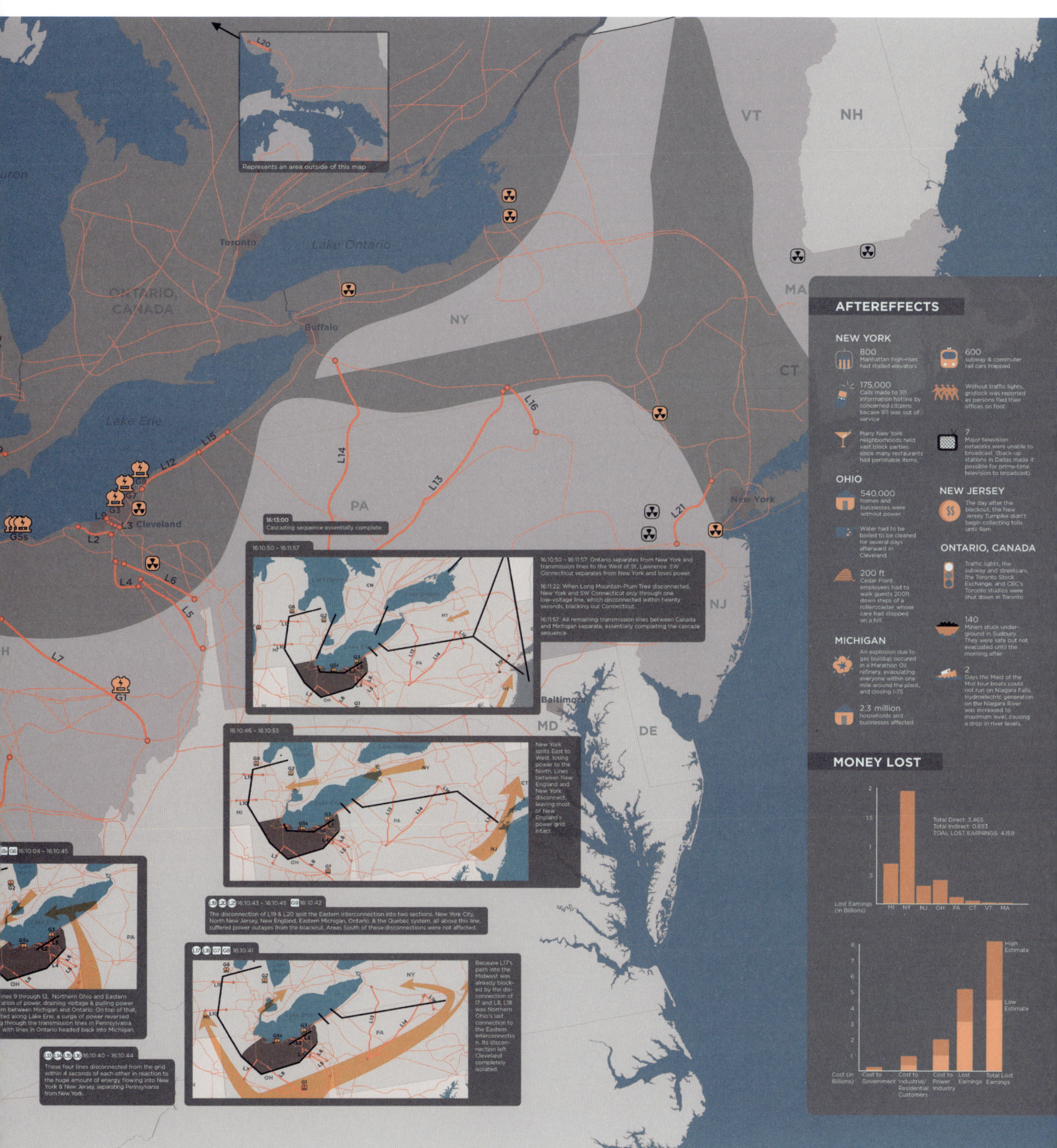

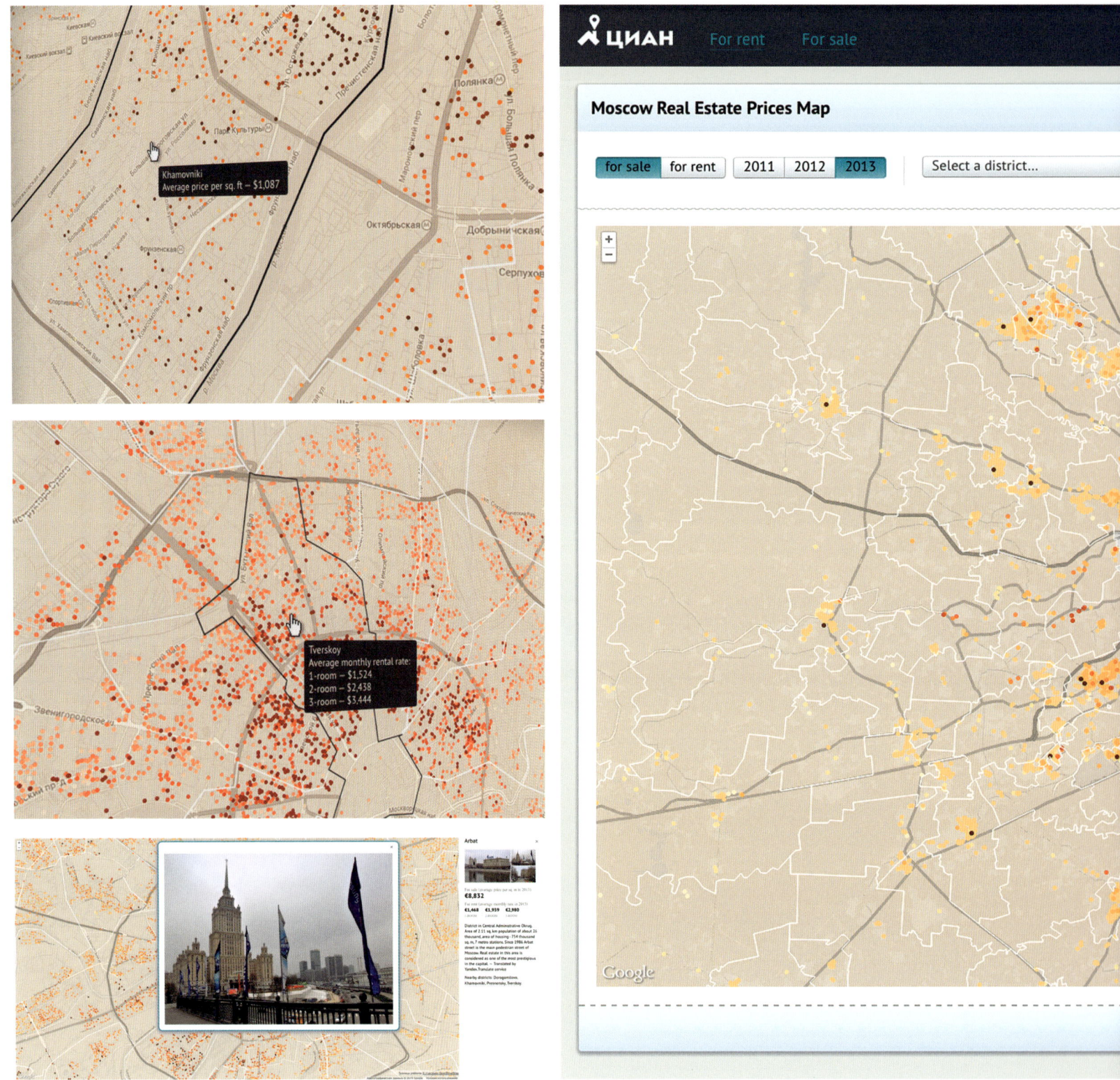

Moscow Real Estate Prices Map

This work aims to be the most detailed and accurate map of property prices in the capital of Russia.
The interactive map shows how the real estate prices in Moscow have changed over the last three years. Properties are visualized on the map by dots which are colored according to their prices. If you select a neighborhood on the map you can find out the average price per square foot in that area. Selecting an area on the map will also load a neighborhood review in the map sidebar.

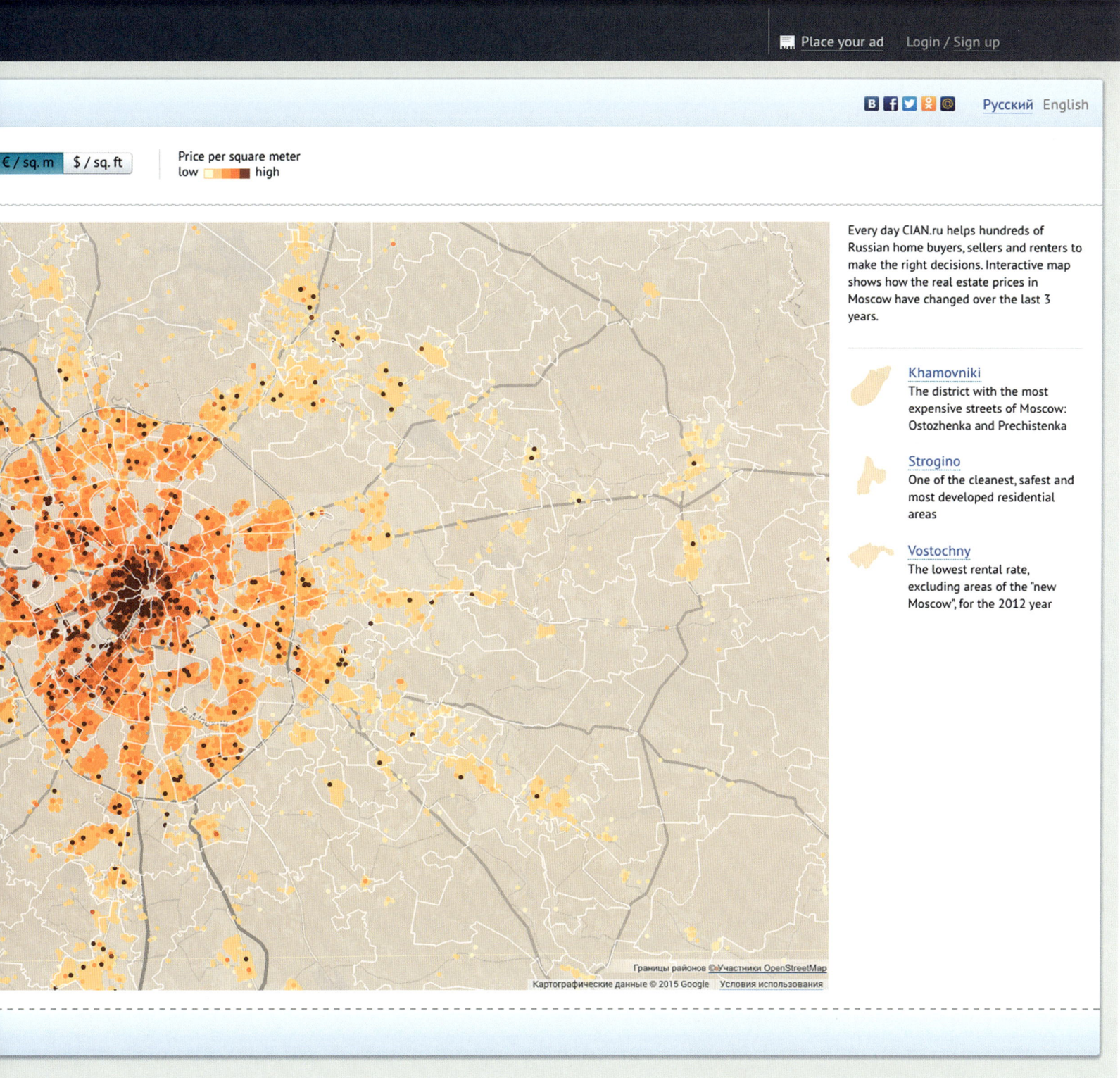

La carte du monde du chocolat

Principaux fabricants de chocolat
Ventes de chocolat et confiseries, en milliards de dollars

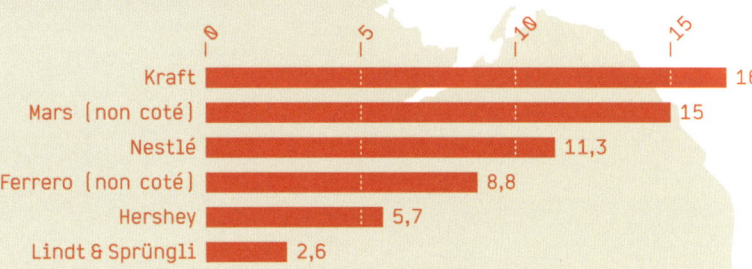

- Kraft — 16,8
- Mars (non coté) — 15
- Nestlé — 11,3
- Ferrero (non coté) — 8,8
- Hershey — 5,7
- Lindt & Sprüngli — 2,6

Principaux broyeurs de fèves
Production, en milliers de tonnes

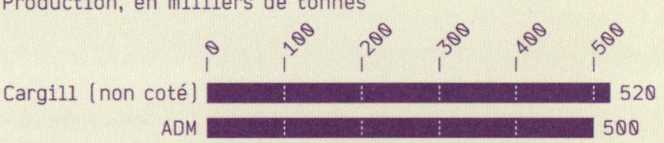

- Cargill (non coté) — 520
- ADM — 500
- Barry Callebaut — 440

Principaux pays producteurs de cacao

Trois pays produisent plus de 65 % du cacao mondial.

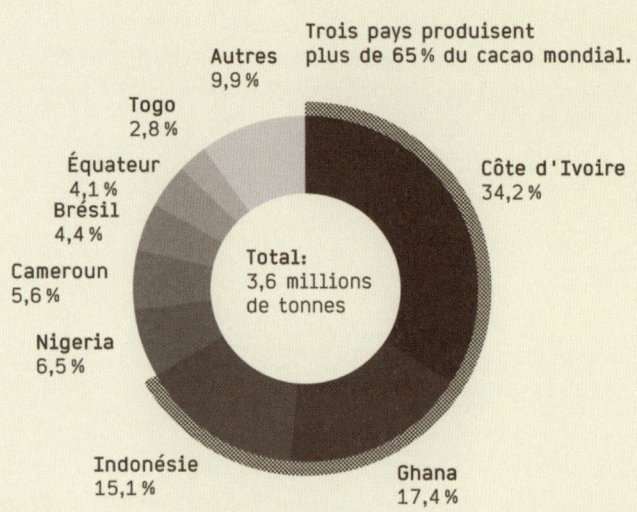

- Côte d'Ivoire 34,2 %
- Ghana 17,4 %
- Indonésie 15,1 %
- Nigeria 6,5 %
- Cameroun 5,6 %
- Brésil 4,4 %
- Équateur 4,1 %
- Togo 2,8 %
- Autres 9,9 %

Total : 3,6 millions de tonnes

Source : International Cacao Organization, **Infographie :** Benjamin Schulte / LargeNetwork

Figure 1

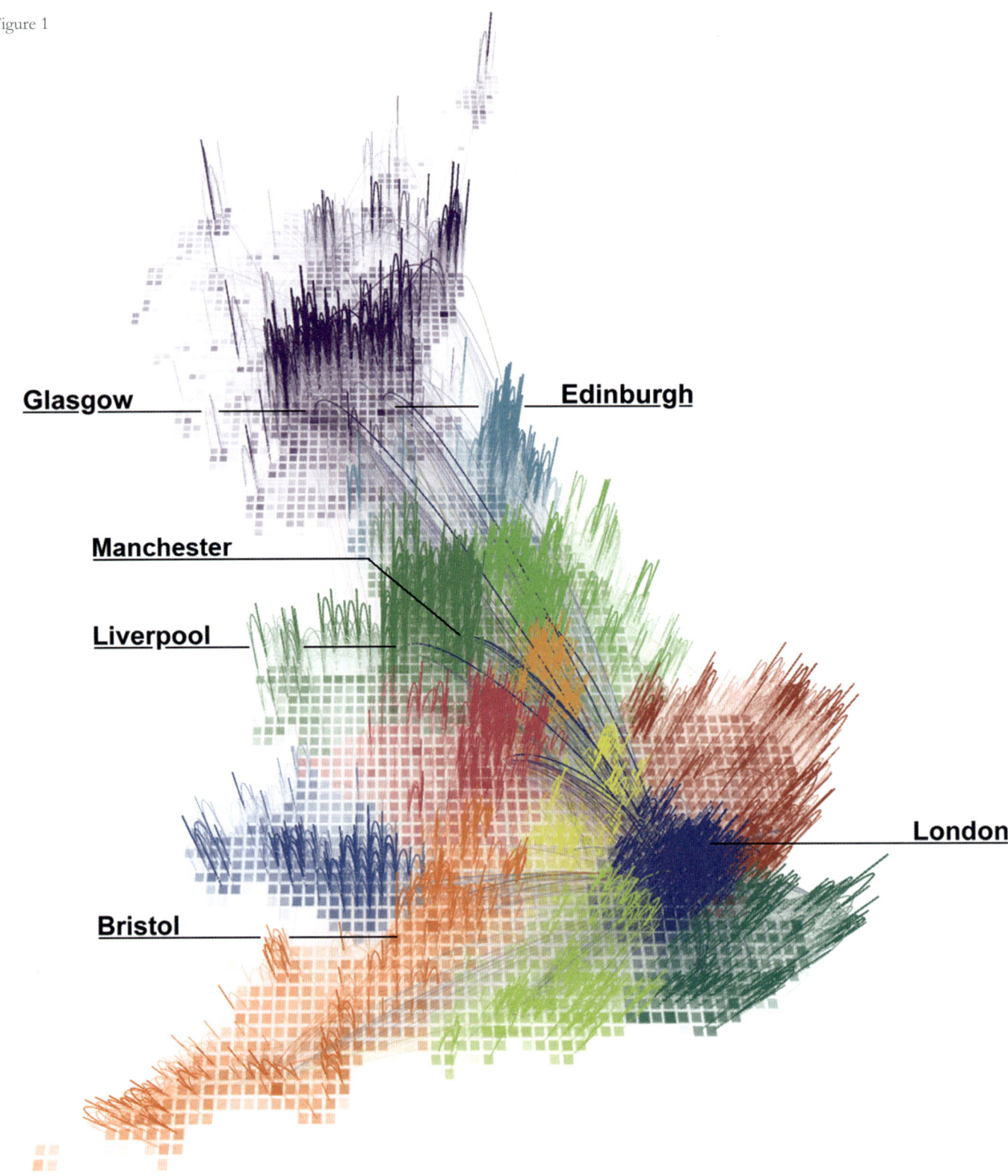

Borderline: Redrawing the Map of Great Britain from a Network of Human Interactions

This map visualizing the geography of talk in Great Britain is created for a paper titled *Borderline: Redrawing the Map of Great Britain from a Network of Human Interactions* which aims to find the answer for the question "do regional boundaries defined by governments respect the more natural ways that people interact across space." Full article available at: *http://dx.plos.org/10.1371/journal.pone.0014248*.

Figure 1. The geography of talk in Great Britain. This figure shows the strongest 80% of links, as measured by total talk time, between areas within Britain. The opacity of each link is proportional to the total call time between two areas and the different colors represent regions identified using network modularity optimisation analysis.

Figure 2

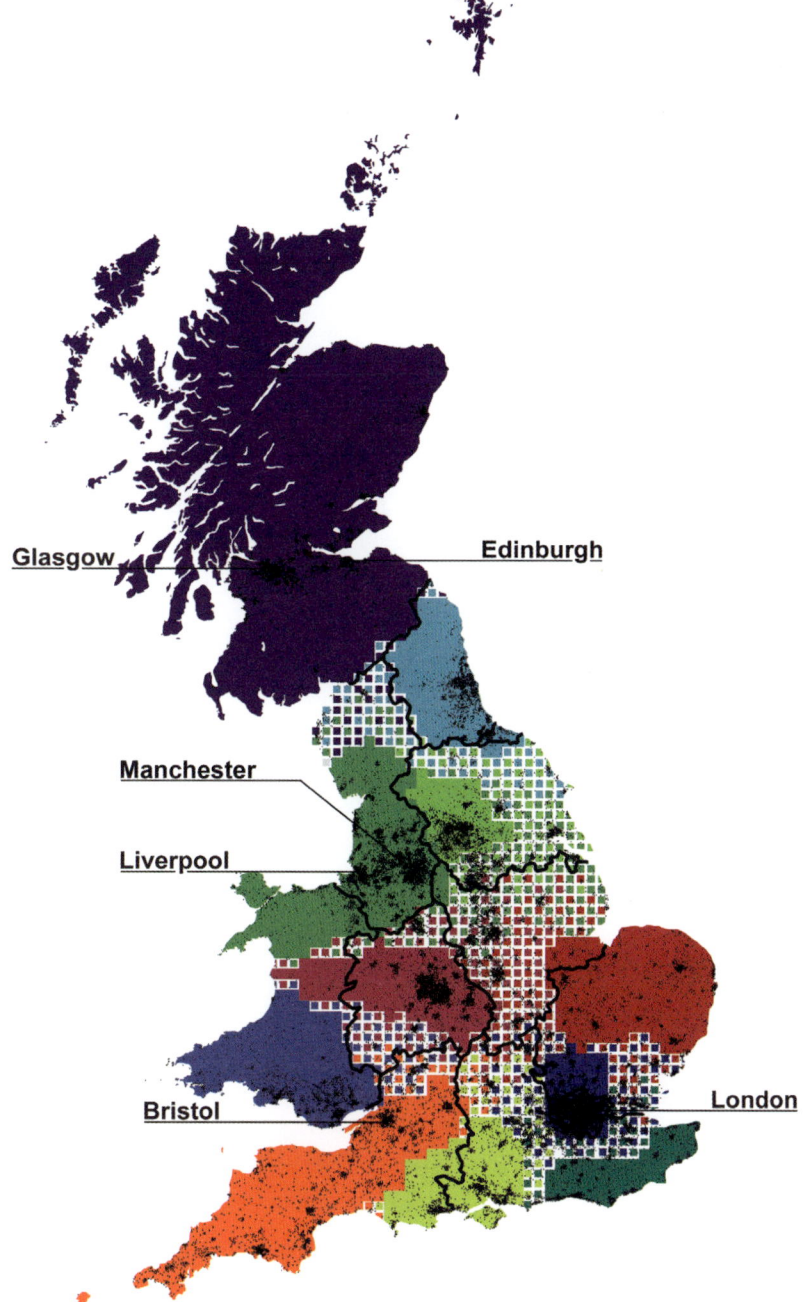

Figure 2. The core regions of Britain. By combining the output from several modularity optimization methods MIT obtains the results shown in this figure. The thick black boundary lines show the official Government Office Regions partitioning together with Scotland and Wales. The black background spots show Britain's towns and cities, some of which are highlighted with a label.

Figure 3

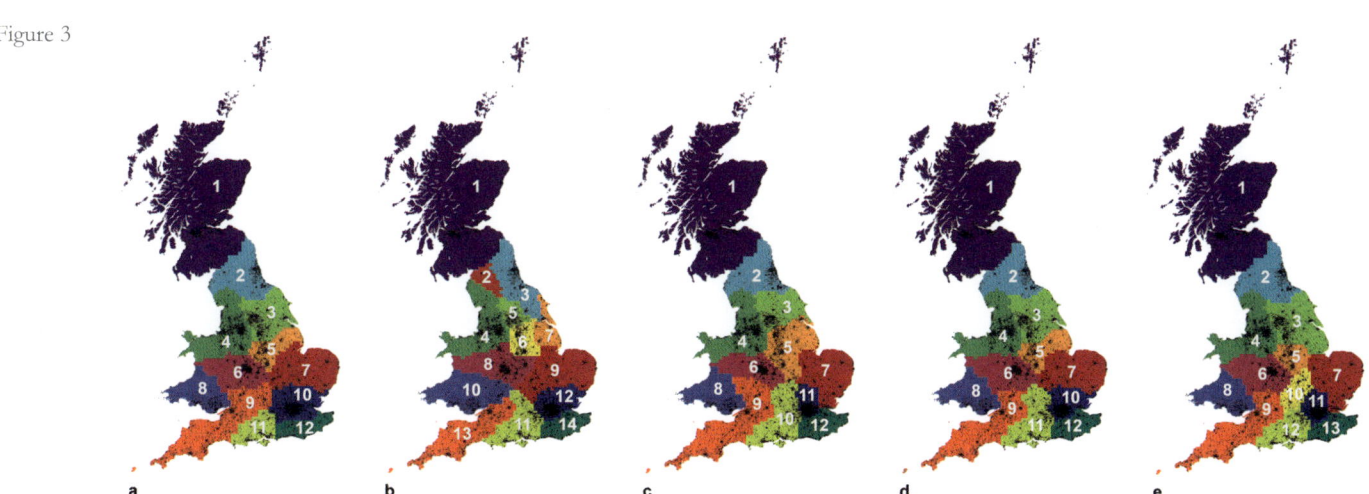

Figure 3. Defining regions through the spectral modularity optimization. Results of five different modularity optimization algorithms.

World Giving Index

This artwork is for Il Corriere della Sera, one of the most important Italian newspapers. The visualization shows the results of the research "World Giving Index", conducted by the Charities Aid Foundation in more than 160 countries, which provides an insight into the scope and nature of giving around the world.

DESIGN Sara Piccolomini   COUNTRY/REGION Italy

WORLD GIVING INDEX ~ DONAZIONI IN DENARO

COME SI LEGGE

La visualizzazione mostra i Paesi più generosi. Il dato è stato ricavato prendendo in esame quante persone hanno fatto almeno una donazione benefica nel periodo precedente all'intervista (2013)

0% — 100% — N.d.

- 6 ISLANDA
- 4 REGNO UNITO
- 4 IRLANDA
- 6 PAESI BASSI
- 52 Italia ~ 28%
- 2 MALTA
- 1 BIRMANIA
- 3 THAILANDIA
- 8 INDONESIA
- 8 AUSTRALIA

Oceano Atlantico · Oceano Indiano

- 2 Malta ~ 78%
- 1 Birmania ~ 91%
- 3 Thailandia ~ 77%

...NO DI PIÙ (TEMPO, SOLDI, BUONE AZIONI)

...ova ...nda	Australia	Malesia	Regno Unito	Sri Lanka	Trinidad e Tobago
...%	56%	55%	55%	54%	54%
	△	▭	△	▭	▭

79° Wgi~Italia
posto in classifica
28% ▽

I CONTINENTI PIÙ GENEROSI ~ OGGI E IERI

Asia	Europa	America	Africa	Oceania
35% 31%	32% 30%	36% 34%	29% 29%	57% 58%
+4%	+2%	+2%	0%	−1%

Legenda

Paese / continente — Wgi 2014 / Wgi 2009-2013 — n% n%

Colore / continente:
- Asia
- Europa
- America
- Africa
- Oceania

- ▭ Wgi invariato rispetto all'anno precedente
- △ Wgi aumentato di almeno 3 punti percentuali rispetto all'anno precedente
- ▽ Wgi diminuito di almeno 3 punti percentuali rispetto all'anno precedente

Incremento percentuale Wgi: +n%

FONTI

Il World giving index è stato pubblicato dalla Charities aid foundation (Caf) nel novembre 2014. La ricerca prende in considerazione tre parametri e registra la risposta degli intervistati sotto forma di percentuale, per poi elaborare un indice generale e classifiche per singole voci

...olontariato) e l'aiuto concreto agli altri. ...sopra: a sinistra, la top ten dell'indice ...tiene conto dei tre parametri del Wgi; ...stra, un confronto tra la generosità dei ...tinenti oggi e quella risultante dalla ...lia delle precedenti rilevazioni (2009-2013)

Blank Map

The map shows the predominant colors of pictures taken in Milan. The source is based on www.panoramio.com. The initial concept for the project was to represent the city examined as if it were a "blank map," i.e. making sure that the shape emerged showing only the predominant colors of the photos taken by locals and tourists. Based on the map data (buildings, roads and parks) from OpenStreetMap, the design team took full advantage of Voronoi Diagram in their experiment and obtained this interesting final result.

The resulting map indicates that the most photographed areas are parks and stations, usually near parks and water areas, so there is a clear predominance of green and blue. Buildings, ground, and shade are common reasons for dark colors in the photos and data.

DESIGN   Valerio Pellegrini & Michele Mauri   COUNTRY/REGION   Italy

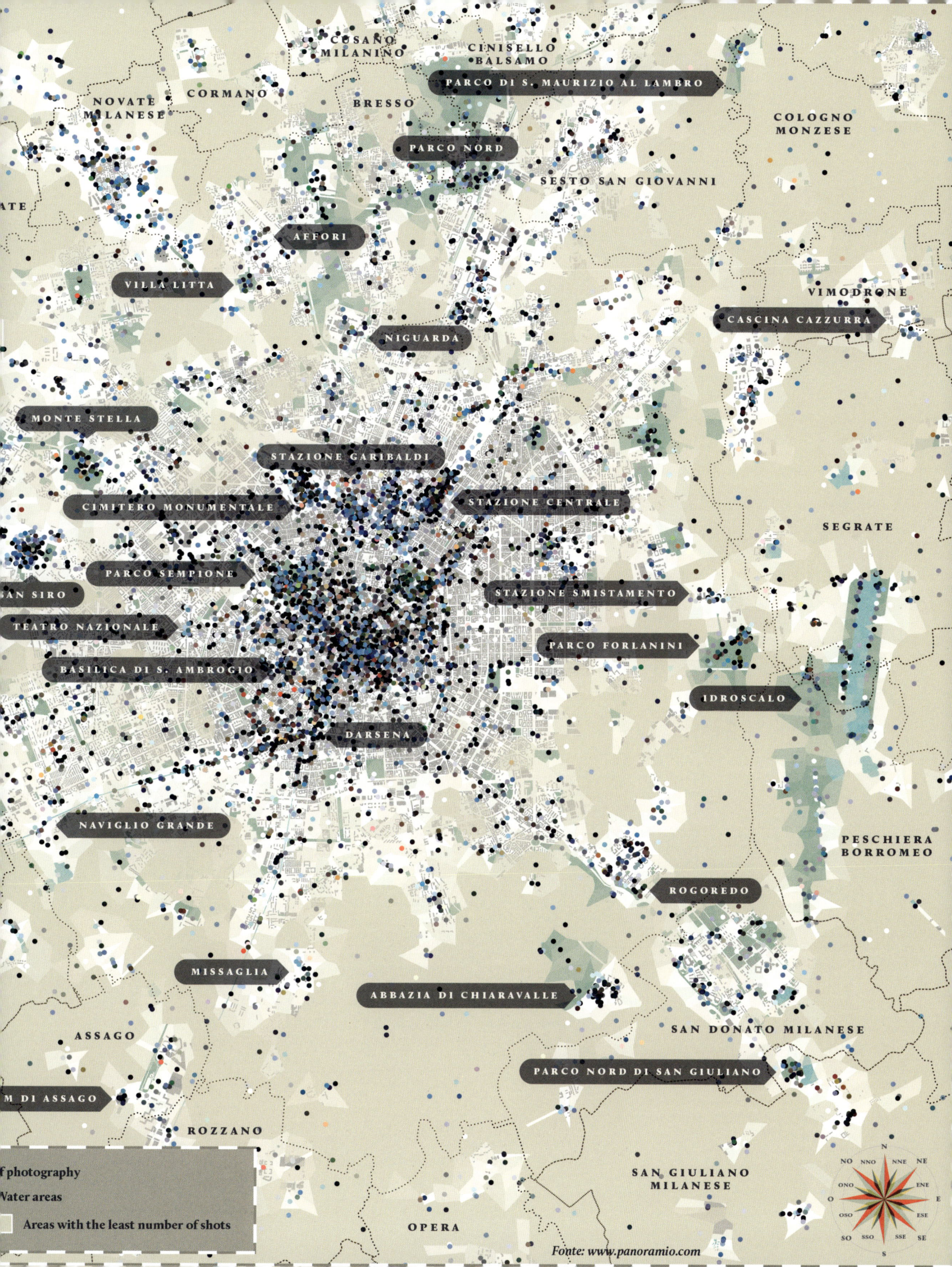

The Future of Thailand

The map was illustrated for Monocle (issue 66.vol.07, September 2013). It's about the Thai government's infrastructure building megaproject in the future. The map shows all the projects that are located in Thailand and Southeast Asia.

The Bermuda Triangle
What lies behind the popular myth

The area between Bermuda and the Bahamas is notorious for its complex navigation conditions. In the mid-20th century, it became known as the Devil's Triangle after a series of ships and aircraft disappeared, which at the time defied natural explanation. The flywheel of mystery, set in motion by the media, turned the Bermuda Triangle into a place surrounded by myths and hypotheses, none of which have earned scientific merit.

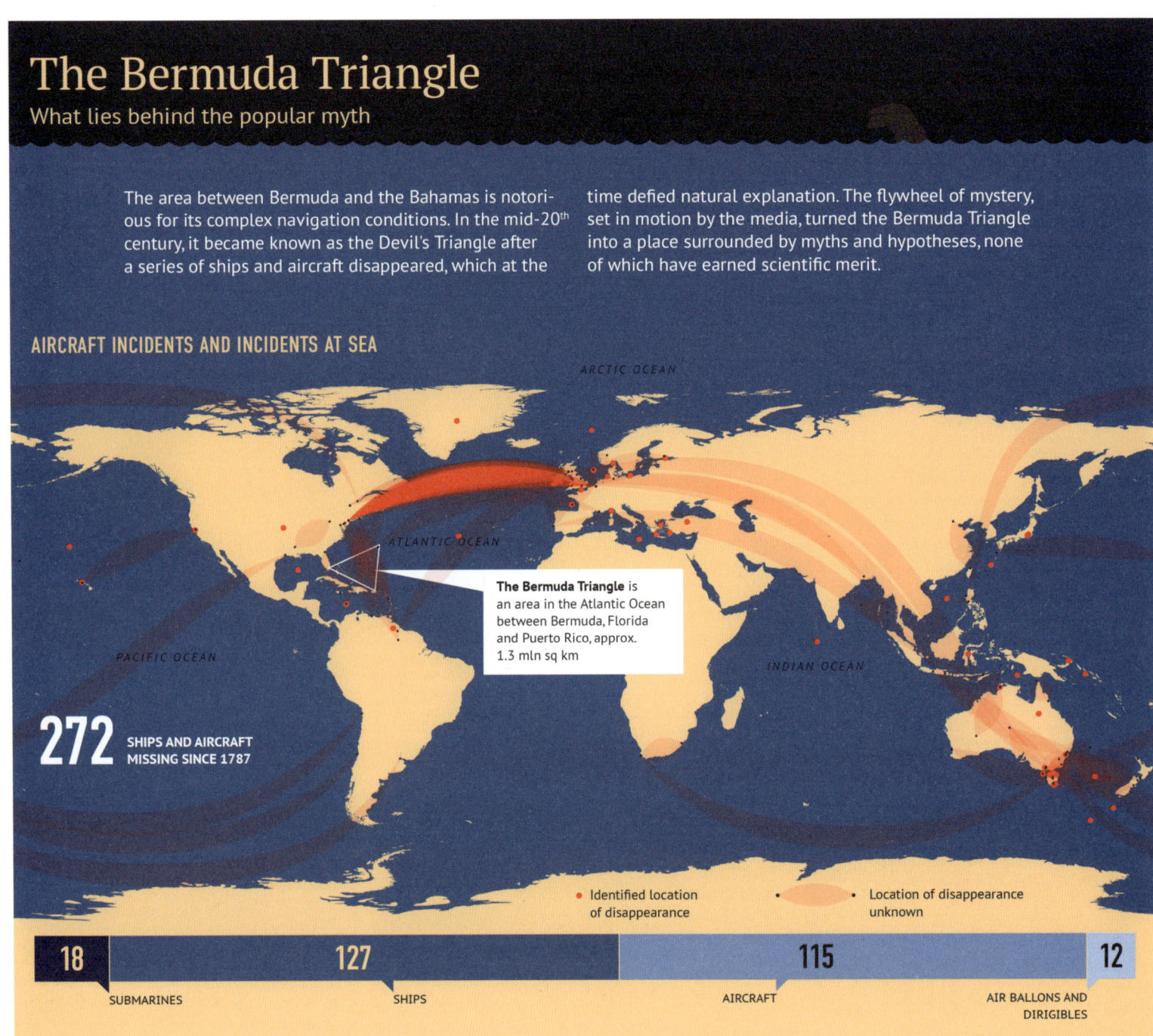

Bermuda Triangle

The area between Bermuda and the Bahamas is notorious for its complex navigation conditions. In the mid-20th century, it became known as the Devil's Triangle after a series of ships and aircraft disappeared, which at the time defied natural explanation. The flywheel of mystery, set in motion by the media, turned the Bermuda Triangle into a place surrounded by myths and hypotheses, none of which have earned scientific merit.

DESIGN Infographic Studio of Rossiya Segodnya International Information Agency   COUNTRY/REGION Russia

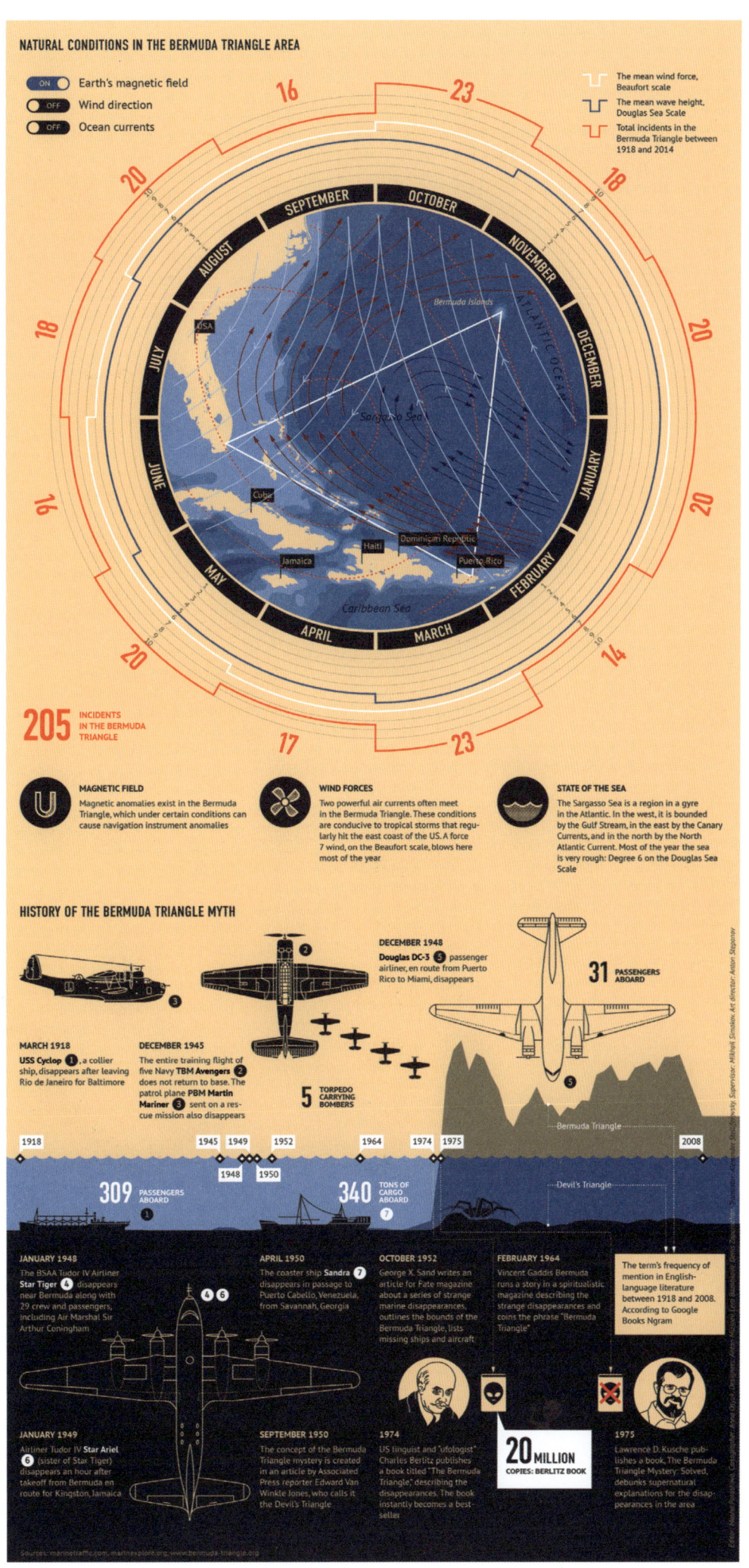

MAPPING HUMAN ACTIVITY

THE DAMNE[D]

THREE GORGES DAM
CHINA

THE YANGTZE IS REVERED FOR ITS ROLE IN PRO[VIDING]
WHO INHABIT ITS BANKS, BUT WITH THE BUILD[ING]
HYDROPOWER PROJECT ALONG THE THREE GOR[GES]
CHINESE PEOPLE IS SEVERELY ABOUT TO CHAN[GE]
SOCIAL, ENVIRONMENTAL AND ECOLOGICAL CON[...]

BEIJING

THE YANGTZE

VOLUME OF WATER
- 529,000 cubic feet per second at the end of the 3 gorges area (Yichang).
- 842,000 Cubic Feet per second at Wuhan.
- 1,100,000 cubic feet per second at it's mouth.
- 3rd volume entering the sea behind the Congo and the Amazon.

CHONGQING

THE CATCHMENT AREA REPRESENTS

- 25% OF CHINA'S ENTIRE CROPLAND
- 33% OF CHINA'S POPULATION, 350 MILLION PEOPLE
- 40% OF CHINA'S PRODUCE OF GRAIN
- 70% OF CHINESE RICE CROP
- 40% OF AGRICULTURAL AND INDUSTRIAL OUTPUT OF CHINA

THE YANGTZE RIVER

The Yangtze river is pinnacle to China's soul and history, great battles have been fought, villages turned into towns and towns into cities. People hid away from the modern world and were able to carry on a way of life unchanged for hundreds of years here.

Tourists come from across the globe to catch a glimpse of what traditional China once was, a sight that can no longer be seen along the banks of the river. The main attraction of the Three Gorges still stand tall above the water level, but somewhat less so towering now.

The middle course effected by the damming stretches 630 miles between Yibin in Sechan and Yichang in Hubei. The river passes through the worlds largest city, with 30 million plus residents in Chongqing, a million of which will also have to be relocated as the dam rises the banks to a level which will allow large ships to come in to port.

YANGTZE FACTS
- **3rd Largest** river in the world.
- More than 700 tributaries drain 1.8 km², approximately **20%** of China's land.
- The growing season within this area lasts for more than **6 months**.

THE W[...]
BECOMI[NG]
NIGH[...]
ENTIRE[...]
ENV[...]
ENDANGER[ED]

37,0[00]
Acre[s]

Singapore
710.2 km³ of land.

LANDSLIDES
There is a significant increase in the risk of landslides in the 3 gorges area of the Yang[tze]

- Since 2004, landslides have forced the relocation of more than **13,000 people** in the county.
- In 2003, just months after the 135 meter water level was achieved, **700m³ feet** (20m³ meters) of rock slid into the Qinggan river adjoined to the Yangtze causing 20 meter high waves that claimed the lives of **14 people**.
- The raise of the water levels to 156 meters in 2006 created **dozens** of landslides that occurred within a **20 mile** distance of the 3 gorges dam.
- Landslides have produced waves as high as **50 meters** (165ft). In July a mountain along a tributary collapsed, dragging **13 farmers** to their deaths and drowning **11 fishermen**.

- In Badong county, November 2007, **4000 cubic yards** (3,050m³) of earth tumbled onto a highway burying a bus and killing at least **30 people**.
- **99 villages** in Miache, 10 miles upstream of the Yangtze, saw land behind them spilt creating **200m wide cracks**. Residents were evacuated to a cave for safety.
- One official said that the shore of the reservoir had collapsed in **91 places** and a total of **36 kilometers** (22 miles) had already caved in.
- Two incidents in May 2009 when **50,000m³** (1,800,000 cu ft) and **20,000m³** (710,000 cu ft) of material plunged into the flooded Wuxia Gorge of the Wu River.

HOW IS A LANDSLIDE CAUSED?
Water seeps into loose soil at the base of rocky cliffs, destabilizing land and making it prone to landslides. The reservoir water level fluctuates, engineers partially drain the reservoir in the summer to accom[modate] flood waters and raise it again at the end of flood season to generate power. This disturbs the land.

According to legend these amazing carvings of mountainous terrain were formed by the goddess Yao Ji would have to redirect the Yangtze around the petrified remains of a dozen dragons she had slain for tormenting peasants. One wonders what Yao Ji would have to say about the mile long Three Gorges dam project now threatening to swallow up the land she once moulded.

YIBIN

The Chinese say If you haven't traveled up the Yangtze, you haven't been anywhere.

The Damned

The Three Gorges Dam in China is the largest hydro-power project in the world. Riley researched the sociological, environmental and ecological effects the dam has had on its surrounding environment along the Yangtze River, and presented the data through this infographic design in an informative, detailed way.

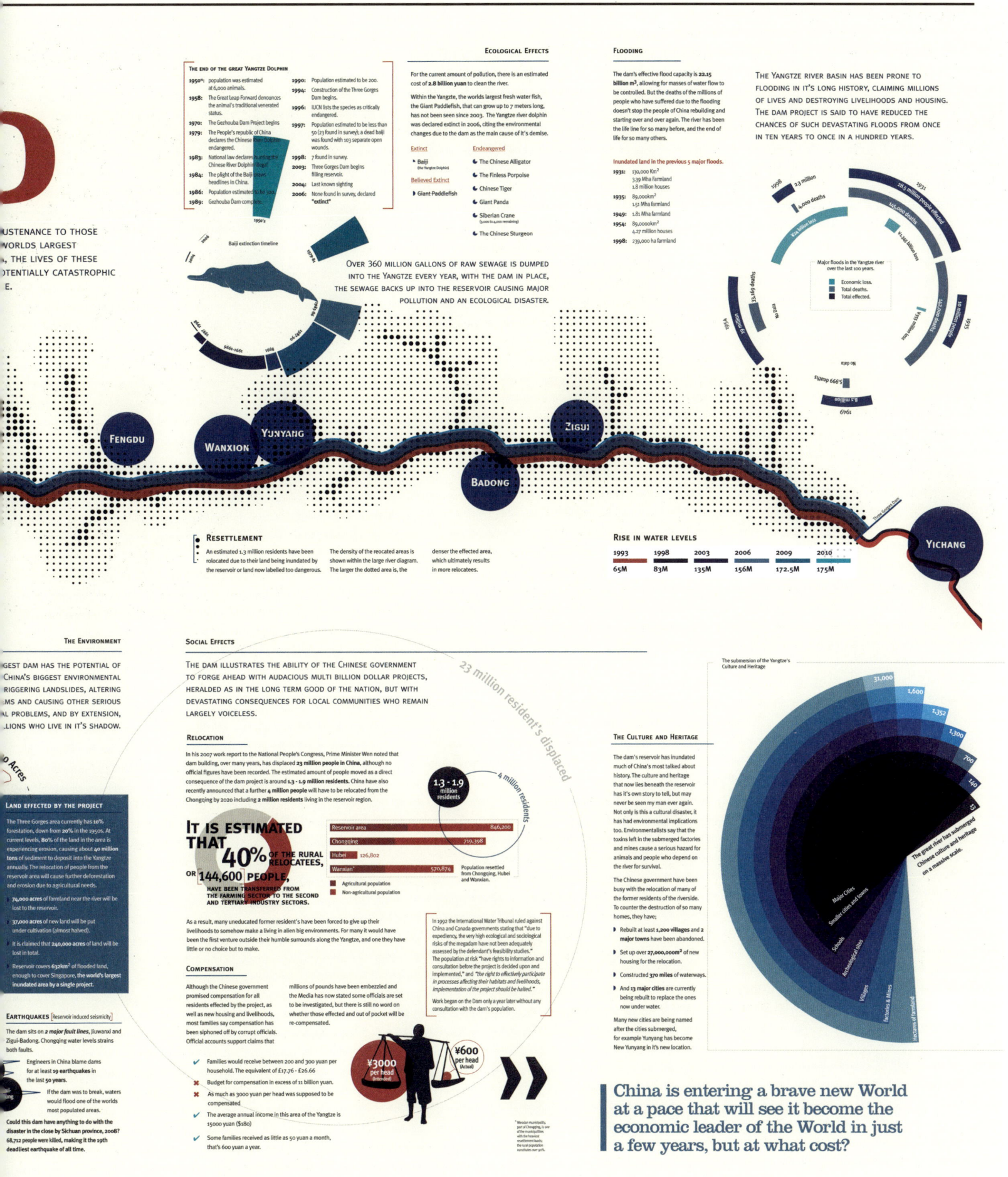

Crossing Paths

As I have grown up, and traveled to more places, I find myself communicating with people who are from, and live, all over the world. I often hear from my friends "I can't keep track of everyone you know, they are everywhere!" This made me realize the paths I have traveled have crossed over other people's paths. Everyone I communicate with is unique in where they are from, where I met them, and where they live currently. I discovered that people who didn't know each other were connected based on locations and by plotting these points of their locations through various stages in their lives, relative to my locations during these phases of my life would be quite interesting.

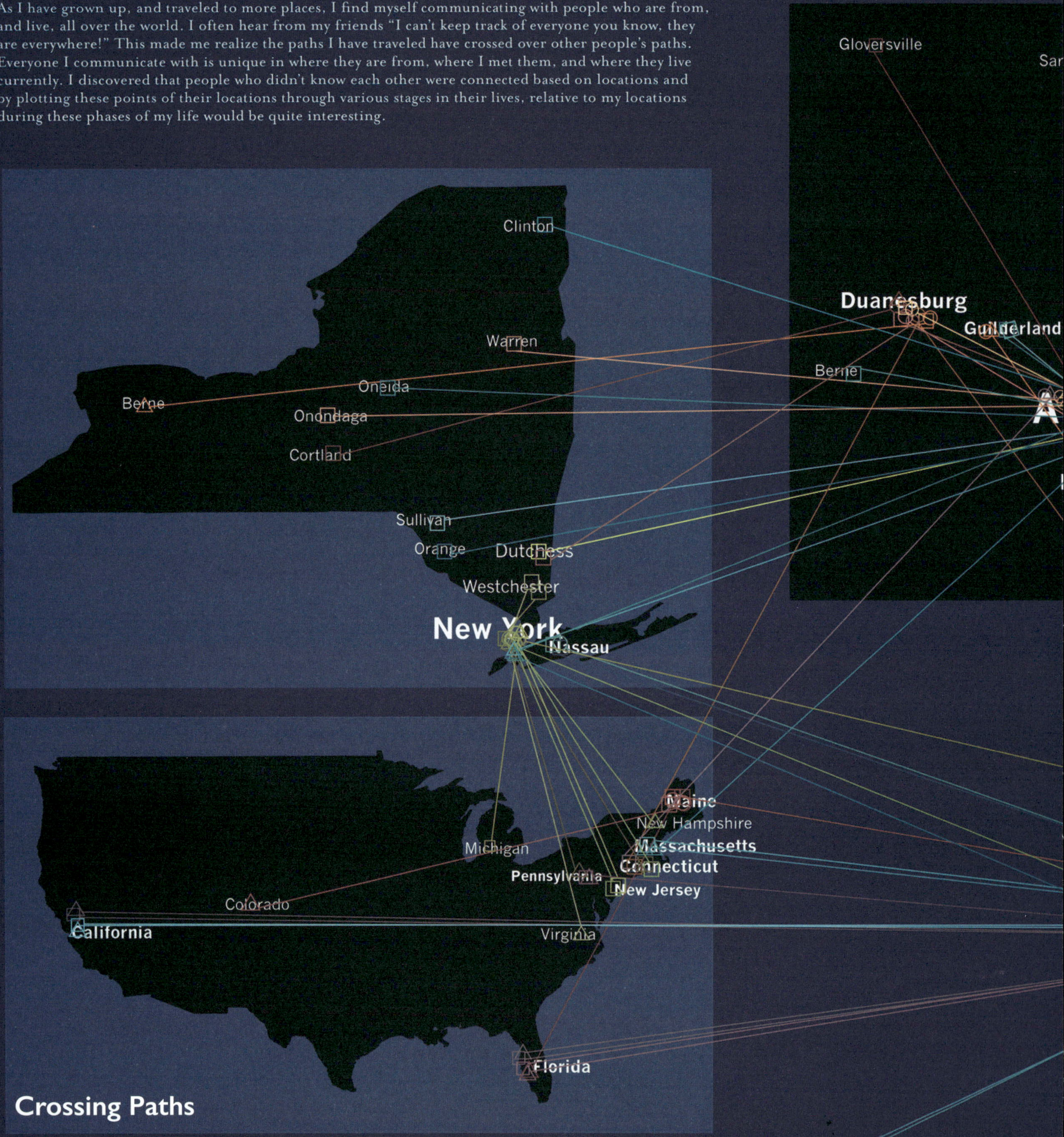

Crossing Paths

Gribbin is an avid traveler and interested in the people and connections she has made throughout the world. This biographical infographic shows the various locations on the map where people Gribbin communicates with are from (square), where she met them (circle), and their current location (triangle). Each person is assigned a specific color, which defines them individually as well as their relationship to the designer.
The map reveals how people were connected to each other geographically even if they had never met. It also shows that certain groups of people tend to remain clustered in the same location while others seem to jump all over the map.

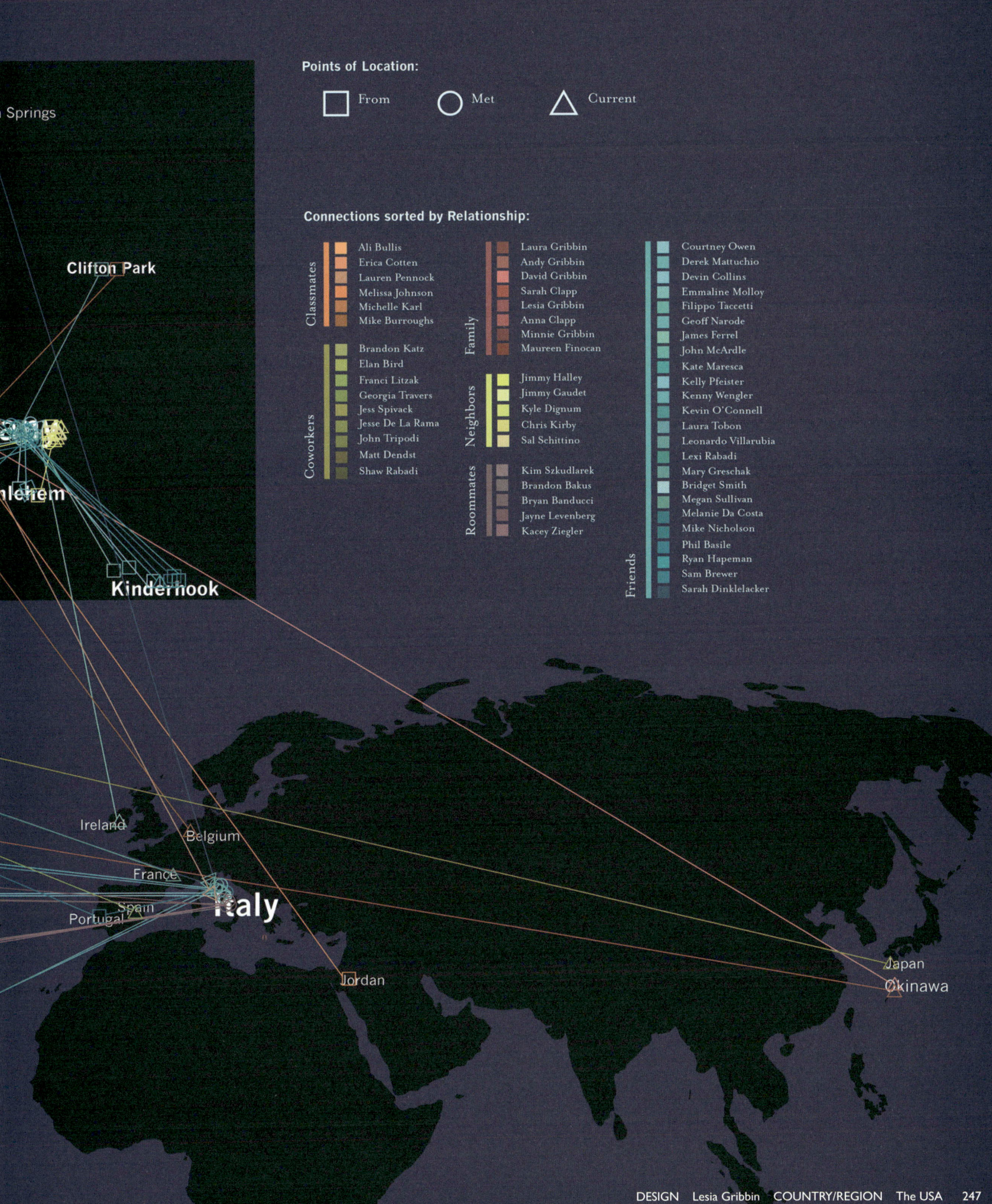

Brochure Design for L'INALCO

The National Institute of Oriental Languages and Civilizations (L'INALCO), says « Languages O'», is a French higher education and research establishment in charge of teaching languages and civilizations, different from those of Western Europe.
This world map is part of a brochure made for L'INALCO. On the back of the brochure, the map presents different languages taught at L'INALCO. To avoid a representation of the world focused on Europe, Graphéine proposed an unusual representation of the map. This perfectly illustrates the promise of openness to the world.

DESIGN Graphéine ILLUSTRATIONS Carole Perret COUNTRY/REGION France

MAPPING HUMAN ACTIVITY

Submap

Submap visualizes locative and time based data on distorted maps. It is a developing mapping series.
The project developed from the idea of drawing a subjective map of Budapest that represents the designers' preferred places or memories in the city. As the places were recognized emotionally 'closer' to the team they would be enlarged where those of less importance would lose focus and become smaller. A web-based tool which can distort maps according to one or more locations was then developed.

DESIGN Kitchen Budapest - Attila Bujdosó, Krisztián Gergely, Dániel Feles, Gáspár Hajdu, László Kiss (sound/music), Tamás Bereznai (graphic) COUNTRY/REGION Hungary

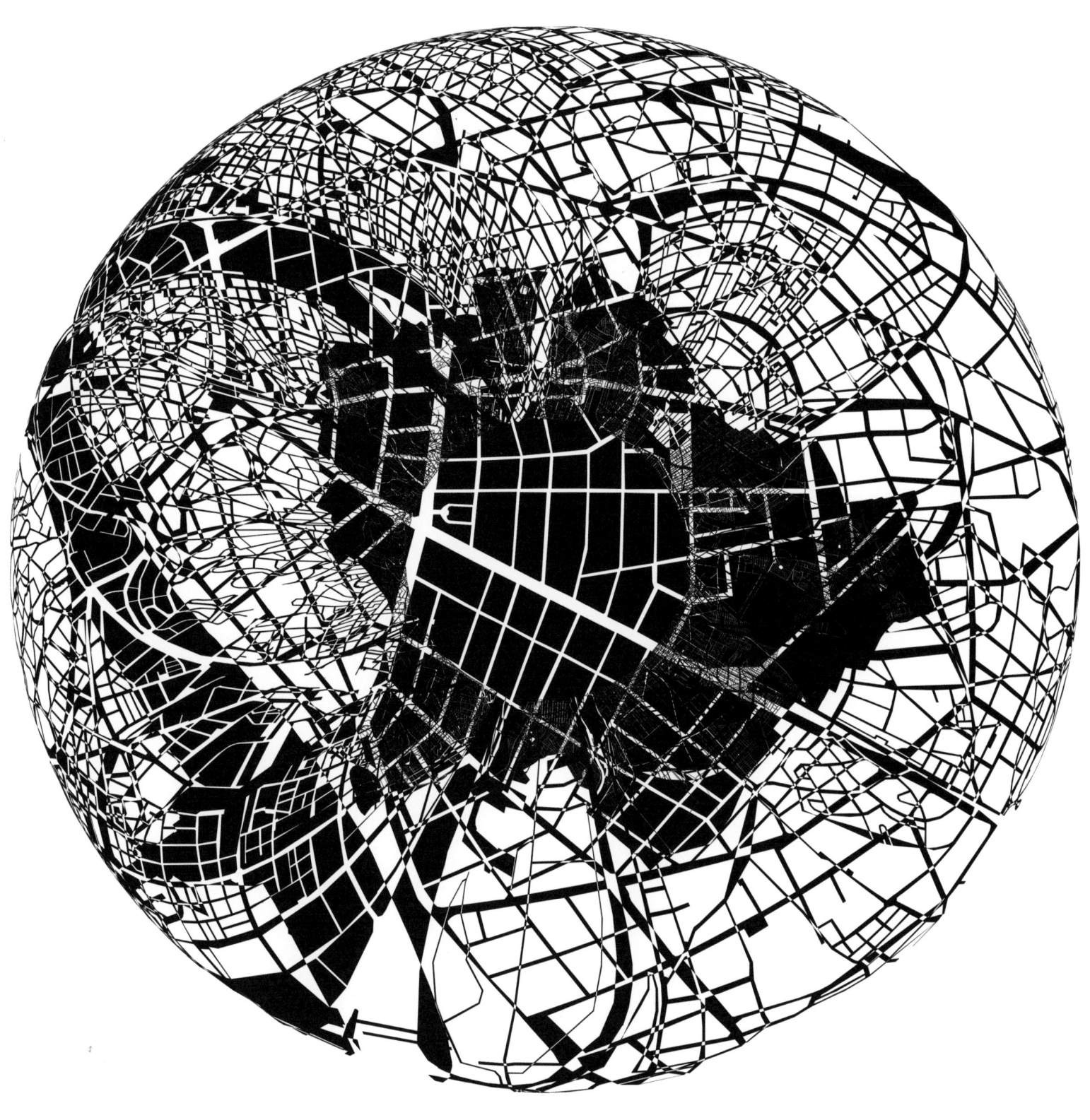

In the first version of SubMap the design team presented a map which shows the city from their point of view. They chose Kitchen Budapest where they all work together as the epicenter of these unique, spherical, perspectival distortions.

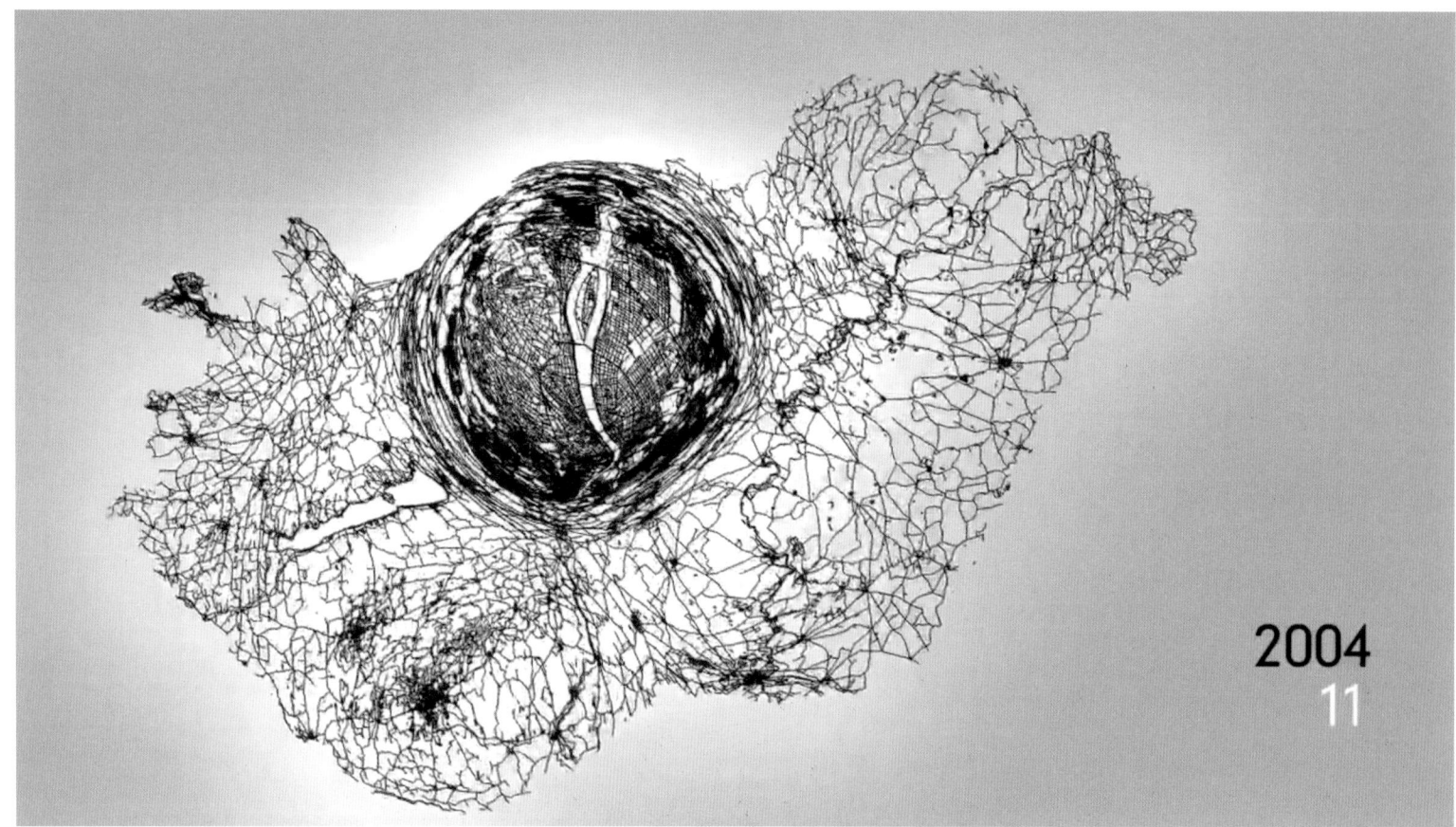

This screenshot is from an updated version of Submap named Ebullition which visualizes and sonificates data pulled from one of the biggest news sites of Hungary, origo.hu. In the 30 fps animation, each frame represents a single day; each second covers a month, starting from December 1998 until October 2010. Whenever a Hungarian city or village is mentioned in any domestic news on origo.hu website, it is translated into a force that dynamically distorts the map of Hungary. The sound follows the visual outcome, creating a generative ever changing drone.

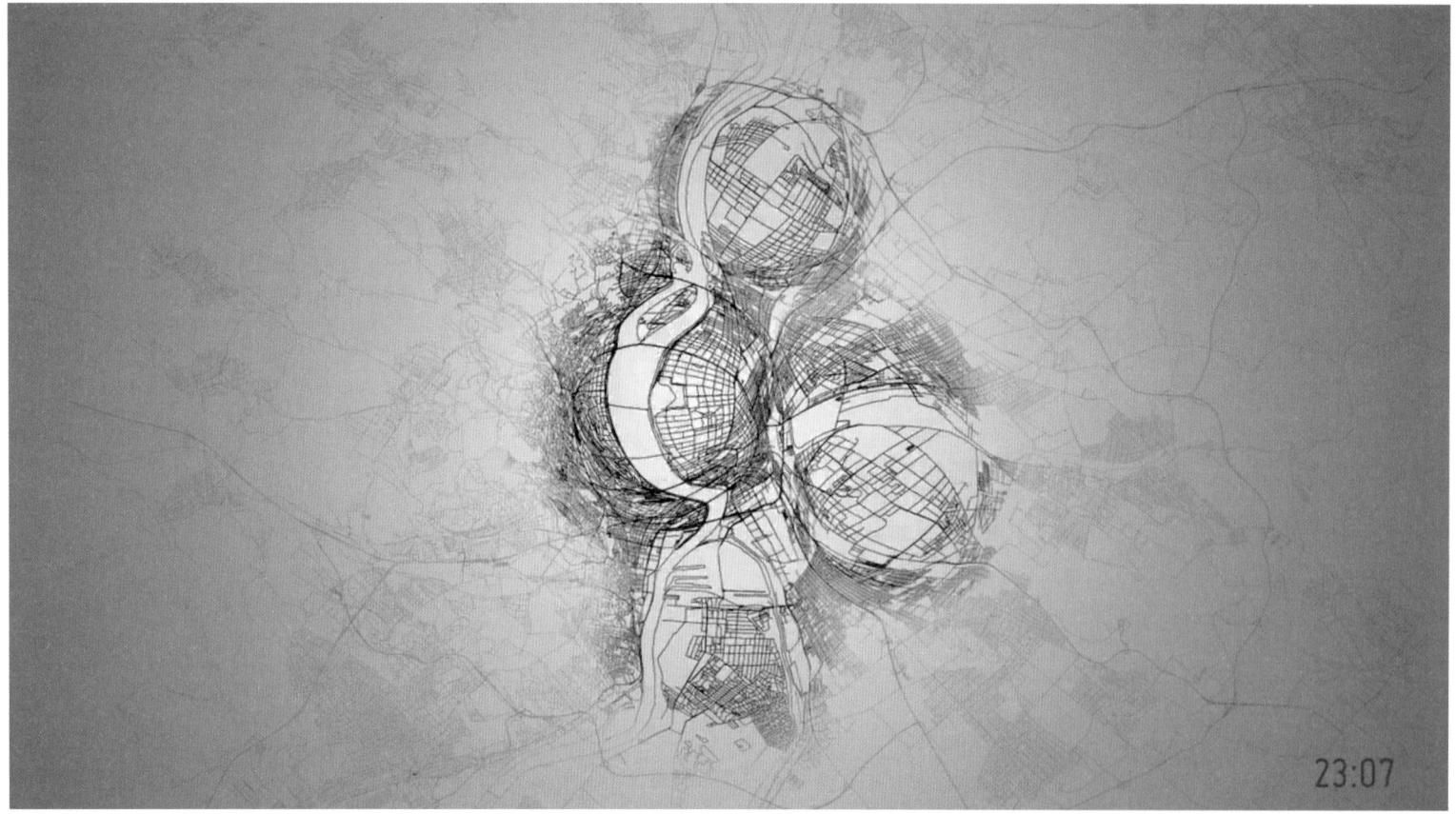

The newest version of Submap was produced by Kitchen Budapest and UrbanCyclr. It untangles the invisible pattern of bike traffic in Budapest. 100,000 kilometres of biking routes collected from individual bikers are overlaid on the city map. All distortions of the map reflect more biking activities in that respective area of the city. A 24 hour map animation reveals the daily biking patterns of a growing community of urban bikers in Budapest. The image above is a screenshot.

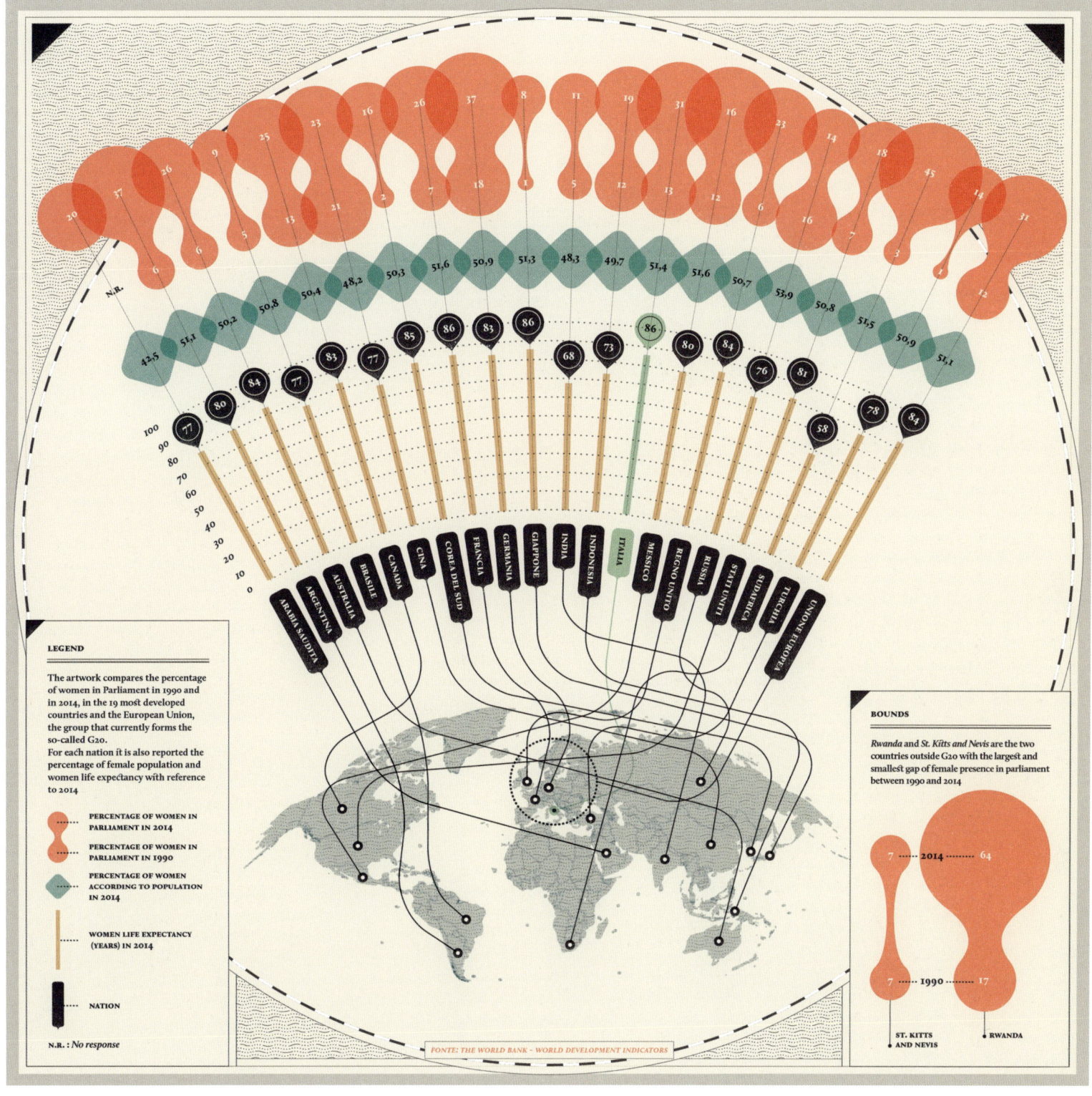

Women in Development

The project shows the presence of women in various positions. Starting from the bottom of the visualization to its top, it represents the average life expectancy of women in 2014, the percentage of women compared to the total population, and finally, the respective percentages of women in parliament in 1990 and 2014 for each of the G20 nations.

DESIGN Valerio Pellegrini COUNTRY/REGION Italy

● INDEX

A

Adrienne Langer
P76
http://www.adriennelanger.com/

Alexey Papulovskiy & Nikolay Guryanov
P158/ P230
http://stat.cian.ru/en

Alfalfa Studio
P59
http://alfalfastudio.com/

Andreas Nilsson
P54
http://andreasnilsson.wix.com/andreas

Antoine Corbineau
P42
http://www.antoinecorbineau.com/

Arianna Di Betta, Veronica Maccari
P226
https://www.ariannadibetta.com

B

Bianca Tschaikner
P28/ P44
http://www.biancatschaikner.com/

C

Camilla Hempleman-Adams
P80
http://camillahempleman.com/

Carrie Winfrey
P222
http://carriewinfrey.com/

Cecilia Della Longa
P198
http://www.ceciliadellalonga.com/

Chaotic Atmospheres
P172
http://www.chaoticatmospheres.com/

Charis Tsevis
P100
http://www.tsevis.com/

Chinapat Yeukprasert
P126
https://www.behance.net/chinapat

Chinapat Yeukprasert & Mandala Studio (Bangkok)
P200
https://www.behance.net/chinapat

Chinapat Yeukprasert & Winks Creatives
P240
https://www.behance.net/chinapat

Ciera Shaver
P46
http://www.cierashaver.com/

Craig Woodward
P20
https://www.behance.net/craigwoodward

CoorsTek, Inc. / Will Manning
P223
http://coorstek.com/

D

Daan Roosegaarde
P144
https://www.studioroosegaarde.net/

Daniel Gaona AKA Mr. Tacho
P12
http://www.mrtacho.com/

Daniel Mason
P194
http://www.daniel-mason.com/

Dénes Sátor
P10
https://www.behance.net/satord

Design Ahoy
P78
http://www.designahoy.com/

Dex from Run For The Hills & Anna Burles
P22
http://runforthehillslondon.com/

Dmitriy Vorontzov
P148/ P150
https://www.behance.net/woodmal

Dom Civiello
P101
http://www.domciviello.com/

Drill Design
P60/ P62
http://www.drill-design.com/

E

Elena Chudoba
P206
https://www.behance.net/elenachudoba

Emma Johnson
P112/ P114/ P115
https://emmaporium.wordpress.com/

F

FIRE / Peter Donnelly Illustration
P122
http://donnellyillustration.com/

Francesco Franchi, Davide Mottes & Danilo Agutoli
P202
https://www.behance.net/davidemottes

G

Gloria Spallanzani
P74
http://www.gloriaspallanzani.com/

Graphéine
P248
http://www.grapheine.com/

H

Hargreaves and Levin
P56
http://www.hargreavesandlevin.com/

I

Infographic Studio of Rossiya Segodnya International Information Agency
P242
http://ria.ru/infografika/

J

Jasmine Desclaux-Salachas
P130/ P160
https://www.facebook.com/CafesCartographiques

Jean Denant
P180
http://www.jeandenant.fr/

Jenni Sparks
P16/ P64/ P116
http://www.jennisparks.com/

Jill Hubley
P174
http://www.jillhubley.com/project/nyctrees

JL Cartography / Jonathan E. Levy
P176/ P182
http://www.jlcartography.com/

Jonathan Hull
P24/ P26/ P96/ P120
https://www.behance.net/jonathanorjack

Ju Hyun Kang
P38
http://hey-jude.kr/

K

Kitchen Budapest - Attila Bujdosó, Krisztián Gergely, Dániel Feles, Gáspár Hajdu, László Kiss (sound/music), Tamás Bereznai (graphic)
P250
http://kibu.hu/

Konstantin Varik
P14
http://varik.ru/

L

LargeNetwork information+design
P232
http://www.largenetwork.com/

Leigh Riley
P215/ P244
http://design.rileigh.co.uk/

Lesia Gribbin
P246
http://www.lesiagribbin.com/

Louise Norman
P110
http://cargocollective.com/louisenormandesign

Lucas Briceno, Bao-Anh Bui & Brendan Gerard
P66
http://www.lbrcn.com/

Luis Dilger
P184
http://www.luisdilger.com/

Luke Johnson & Christiane Holzheid & Erin Ellis
P86
http://cargocollective.com/erinellis/Curiosities-Map

M

Marc Inzon
P154
http://www.marcinzon.com/

Marco Bernardi, Federica Fragapane & Francesco Majno
P134
https://www.behance.net/FedericaFragapane

Martina Sikiric
P142
https://www.behance.net/msart

Massimiliano Mauro
P218
http://massimilianomauro.com/

Matthew Cusick
P128/ P136
http://mattcusick.com/

Matthew Picton
P68
http://matthewpicton.com/

Michael Tompsett
P82
http://michael-tompsett.artistwebsites.com/

Mike Hall
P102/ P104/ P106/ P108
http://www.thisismikehall.com/

MIT SENSEable City Lab
P234
http://senseable.mit.edu/

MYDM co., ltd
P146
http://www.mydm.me/home.html

N

N+studio/Ningchao Lai
P228
http://ningchaolai.com/

Nils-Petter Ekwall
P155
http://www.nilspetter.se/

NOMO Design/Jerome Daksiewicz
P164
http://nomodesign.com

O

Owi Liunic
P118
https://www.behance.net/owiliunic

P

Pablo Espinosa
P90
http://sugacyan.com/

Paul Mathisen
P18
http://www.paulmathisen.com/

Paul Mullen
P92
http://www.paulmullen.co.uk/

Paulina Urbanska
P124
https://www.behance.net/paulinaurbanska

Philip Johnson
P208
http://www.philipjohnson.com/

R

Raushan Sultanov
P98
https://www.behance.net/Rushavel

relajaelcoco studio
P165
http://www.relajaelcoco.com/

relajaelcoco studio/ Pablo Galeano & Francesco Furno
P196/ P212
http://www.relajaelcoco.com/

Ren Ri
P188
http://weibo.com/u/2757365410

Rhiannon Fox
P224
http://www.rhiannonfox.com/

Rod Hunt
P216/ P220
http://www.rodhunt.com/

Ruth Rowland & Seven Publishing Group
P152
http://ruthrowland.co.uk/

S

Sam Williams Studio
P166
https://samwstudio.wordpress.com

Sara Drake
P48
http://www.saradrake.com/

Sara Piccolomini
P236
http://www.sarapiccolomini.com/

Sidney Jablonski
P168
https://www.behance.net/SidJ

Steph Marshall
P72
http://www.stephmarshall.co.uk/

Studio MaMa
P156
http://studiomama.ph/

Sung H. Cho
P170
http://www.sunghcho.com/

Suwanna Ruayrinsaowarot
P162
http://www.suwannar.com/

T

Tilo Richter
P53/ P94
http://www.newyorktilotokyo.com/

Tomasz Kowal
P30/ P32/ P36
http://www.tomaszkowal.pl/

V

Valerio Pellegrini
P253
https://www.behance.net/valeriopellegrini

Valerio Pellegrini & Michele Mauri
P238
https://www.behance.net/valeriopellegrini

Vincent Meertens
P192/ P214
http://www.vincentmeertens.com

Vladislava Savic
P210
https://www.behance.net/VladislavaSavic

W

Wu, Jui-Che
P52
https://www.behance.net/wujuiche

Z

Z.CHENG LEO
P50
http://www.liaozcheng.com/

ACKNOWLEDGEMENTS

We would like to thank all the designers and contributors who have been involved in the production of this book; their contributions have been indispensable to its creation. We would also like to express our gratitude to all the producers for their invaluable opinions and assistance throughout this project. And to the many others whose names are not credited but have made specific input in this book, we thank you for your continuous support.

FUTURE COOPERATIONS: If you wish to participate in SendPoints' future projects and publications, please send your website or portfolio to editor01@sendpoints.cn